Selected Problems in
PHYSICS
with Answers

M. P. Shaskol'skaya
and I. A. El'tsin

Translated by
W. J. F. Reynolds

Dover Publications, Inc.
Mineola, New York

Bibliographical Note

This Dover edition, first published in 2013, is an unabridged republication of the work originally published in 1963 by Pergamon Press, Oxford, and the Macmillan Company, New York. It was originally published in Russian in 1959 by Fitmatgiz, Moscow, as *Sbornik Izbrannykh Zadach Po Fizike.* The English translation was edited by F. Castle, Senior Lecturer in Physics, Brighton Training College, UK.

International Standard Book Number

ISBN-13: 978-0-486-49993-2
ISBN-10: 0-486-49993-6

Manufactured in the United States by LSC Communications
49993602   2017
www.doverpublications.com

Contents

		PAGE	
TRANSLATION EDITOR'S FOREWORD		vii	
AUTHORS' FOREWORD		viii	
		Problems	*Solutions*
I.	Kinematics	3	85
II.	The dynamics of motion in a straight line	9	96
III.	Statics	18	117
IV.	Work: Power: Energy: The law of conservation of momentum: The law of conservation of energy	23	128
V.	The dynamics of motion in a circle	30	146
VI.	The universal theory of gravitation	34	155
VII.	Oscillation: Waves: Sound	36	160
VIII.	The mechanics of liquids and gases	41	172
IX.	Heat and capillary phenomena	56	198
X.	Electricity	61	211
XI.	Optics	76	233

Translation Editor's Foreword

To understand anything is never just a matter of asking questions. It is much more a matter of establishing priorities, then asking the right questions in the right order.

Some attempt has been made to do this in *Selected Problems in Physics*. Some questions will be familiar to those concerned with teaching the subject, others will not. But by asking, and answering as fully as possible, questions demanding a certain resourcefulness on the part of the reader the authors have hoped to produce an essentially readable book which will make a direct contribution to his knowledge. As such it was considered suitable for inclusion in the Commonwealth and International Library.

Particular attention has been directed to the comprehensive nature of the answers so that the book should be useful to a considerable range of persons concerned with some aspects of applied mathematics and with physics. It is hoped that these in their later school days or in the early stages of a technical college course will find the answers to some of the questions they are asking and being asked; and to some they are not.

F. Castle

Authors' Foreword

THE present collection of problems is a further development and revision of our book *Selected Physics Problems*, which was published in 1949 and was soon sold out. The basis of our earlier book was formed by problems set over a number of years in the "Olympic" examinations set in Physics to schoolchildren by the Physics Faculty of the Lomonosov State University in Moscow. A large number of teachers and a number of the students of the Physics Faculty of the Moscow State University took part in composing and selecting the "Olympic" problems.

In its new, revised form our book has been augmented by a large number of new problems; only a few of them are "Olympic" ones. We have tried to keep the previous style of problem, avoiding the commonplace ones and choosing only those which demand more resourcefulness and inventiveness than do normal school-problems.

Most of these problems can be solved from the knowledge of physics acquired in school; but we have not felt ourselves confined within the limits of the secondary-school syllabus, but have counted on pupils who have an interest in physics and are widening their knowledge by independent reading. The solution of such problems, or even an attentive analysis of the solutions given, should help schoolchildren to learn to apply their knowledge when grappling with concrete problems.

We hope that the present book will be of use, not only to schoolchildren, but also to their teachers and to students in technical and other advanced colleges.

THE AUTHORS

Problems

I. Kinematics

1. Two passengers with stop-watches decide to measure the speed of a train: one by the click of the wheels passing over the junctions between rails (knowing that the length of each rail is 10 m), the other by the number of telegraph-poles passing the window (knowing that the distance between them is 50 m). The first passenger starts his stop-watch on the first click and stops it on the 156th. He finds that 3 min have passed. The second passenger starts his stop-watch when the first telegraph-pole appears in the window and stops it when the 32nd pole appears. He too finds 3 min have passed. The first passenger calculates that the train's speed is 31.2 km/hr, and the second that it is 32 km/hr. Which of them has made a mistake and how? What is the real speed of the train?

2. The journey from port A to port B lasts exactly 12 days. Every midday a steamer sets out from A for B and another from B for A. How many steamers does each boat meet in the open sea?

3. What exposure should be given to a photograph of a car

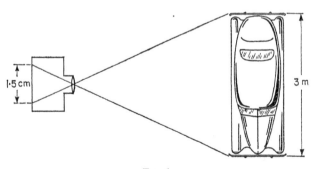

Fig. 1

moving at a speed of 36 km/hr (Fig. 1), so that the image on the negative should not be blurred—taking that for this the image should not move more than 0·1 mm? The length of the car is 3 m, and the resulting image is 1·5 cm long.

4. A car travels a distance from A to B at a speed of 40 km/hr (v_1) and returns at a speed of 30 km/hr (v_2). What is its average speed for the whole journey?

5. A boy is throwing balls into the air, throwing one whenever the previous one is at its highest point. How high do the balls rise if he throws twice a second?

6. Two stones fall down a shaft, the second one beginning its fall 1 sec after the first. Find the second stone's motion in relation to that of the first. Ignore air-resistance.

7. Two planes are flying at the same speed of 200 m/sec in opposite directions. A machine-gun mounted in one plane fires at the other at right angles to their line of flight (Fig. 2). How far apart will the bullet-holes made in the side of the second plane be, if the machine-gun fires 900 rounds per minute? What role does air-resistance play in this?

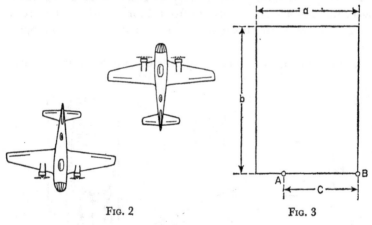

FIG. 2   FIG. 3

8. A billiards-ball is at point A on a billiards-table whose dimensions are given in Fig. 3. At what angle should the ball be struck so that it should rebound from two cushions and go into pocket *B*? Assume that in striking the cushion, the ball's direction of motion changes according to the law of reflection of light from a mirror, i.e. the angle of reflection equals the angle of incidence.

9. You are given three billiards-tables of different lengths and the same width. Balls are struck simultaneously from the edge of one of the long sides of each table (Fig. 4) with velocities which are equal in direction and magnitude. Is it possible that these balls should not return to the side from which they started at exactly the same moment?

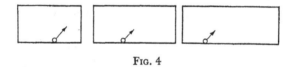

FIG. 4

10. A bucket is left out in the rain. Will the speed at which the bucket is filled with water be altered if a wind starts to blow?

11. A tube is mounted on a trolley which moves uniformly in a horizontal plane (Fig. 5). At what angle to the horizontal should the tube be inclined so that a drop of rain, falling perpendicularly, should reach the bottom of the tube without touching its sides? The raindrop's rate of fall, v_1 is 60 m/sec (which does not alter, thanks to the effect of air-resistance). The speed of the trolley, v_2 is 20 m/sec.

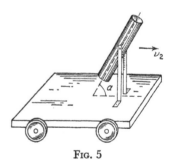

FIG. 5

12. To find the speed of a river's current, a boatman decides to carry out the following experiment. He lowers a wooden bucket into the water and himself sets off downstream, rowing. After 40 min he reaches a point *A*, 1 km from his starting-point and turns back. He picks up the bucket, turns round again and, rowing downstream once more, reaches A for the second time 24 min later. What is the speed of the current, assuming that the speeds of both current and boat are constant, and also that no time is wasted on turning round? How long does the oarsman spend on rowing upstream to meet the bucket? What is the boat's speed relative to the water?

13. Why is it that when a car is moving forward on a cinema-screen, the wheels often appear to be turning backwards?

14. If a disk with one or more holes pierced in it is placed in front of a beam of light which lights up drops of water falling one after the other and the disk is then rotated, the beam will light up the drops intermittently. The number of flashes will depend on the speed of rotation of the disk and on the number of holes in it. This method of illumination is called stroboscopic; it permits periodic phenomena to be observed which are taking place at such a speed that it is impossible to observe them with ordinary lighting. If the number of revolutions of the stroboscope is so chosen that in the time between two flashes the drops have time to move a distance equal to the distance between successive drops, then the drops will appear to be stationary. Find the number of revolutions of the disk necessary for this, if the disk has two holes, if the distance between the drops, $s = 2$ cm, and the height from which the drops fall is $h = 22 \cdot 5$ cm.

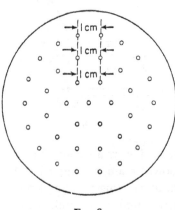

Fig. 6

15. A disk with holes pierced in it at distances of 1 cm along the circumferences of concentric circles (Fig. 6), is lit from behind by a lamp. The disk rotates with a speed of 30 rev/min. At what distance from the centre of the disk shall we see a continuous circle of light? The human eye does not distinguish between alternating periods of light and dark, if their frequency is greater than 16 to the second.

16. A hoop of radius R rolls without slipping along a horizontal plane with constant speed v. What is the acceleration of different points on the hoop's circumference?

17. A man holds one end of a plank, while the other end rests on a drum (Fig. 7). The plank is horizontal. Then the man moves the plank forward, making the drum roll without slipping

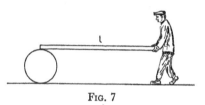

Fig. 7

along a horizontal plane; no slipping takes place either between plank and drum. How far must the man move before reaching the drum, if the plank is of length l?

18. A hoop is thrown on to a rough horizontal plane with a linear speed v. At the same time the hoop is given a rotatory movement in a direction such that the hoop will roll in the direction of the throw (Fig. 8). What angular velocity, ω, will make the hoop roll along the plane without slipping, if the hoop's radius is R?

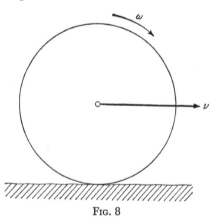

Fig. 8

19. When a wheel is in motion, the upper spokes often seem to merge, while the lower ones are distinct. Why is this?

20. At what speed should an aeroplane fly horizontally on the latitude of Leningrad (60°) so that the pilot should be able to see the sun always in the south? The radius of the earth is 6300 km.

21. Two men decide to fight a duel with revolvers in unusual circumstances: they are to fire while standing on a roundabout of radius R, which is turning with an angular velocity of ω. The first duellist stands at the centre O of the roundabout, the second at its edge. How should they each aim so as to hit his opponent? Which is in the more favourable position? Assume that the first duellist's bullet is fired from O at a velocity v.

II. The Dynamics of Motion in a Straight Line

22. A bomb is dropped from an aeroplane flying horizontally at a constant speed. Where will the aeroplane be when the bomb hits the ground?

FIG. 9

23. The barrel of a gun and the centre of a target, hung from a thread, are in a horizontal straight line (Fig. 9). Will the bullet hit the target if the thread breaks and the target begins to fall freely at the moment the bullet leaves the muzzle? Assume that there is no air-resistance.

24. Which raindrops fall faster, big ones or little ones? Why?

25. Two spheres of the same radius and the same material fall through the air from the same height; one sphere is solid, the other is hollow. Which will fall faster?

26. A tube in the shape of a rhombus with rounded corners is placed in a vertical plane as shown in Fig. 10. A ball is allowed to roll inside the tube along sides AB and BC, and then allowed to

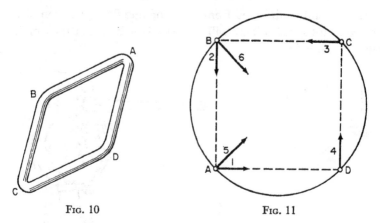

Fig. 10   Fig. 11

roll along sides AD and DC. In which case will it roll faster? The length of the rhombus's side is A.

27. A load of mass m begins to slip without friction down the inclined face of a wedge lying on a horizontal plane surface; there is no friction either between wedge and plane. The mass of the wedge is M, the angle of inclination of the wedge's top surface with the horizontal is α. Find the acceleration of the load and of the wedge relative to the plane, the force exerted by the load on the wedge and by the wedge on the plane.

28. In Fig. 11 is shown a thin ring of radius R. Equal forces are acting at points A, B, C, D, which are the vertices of an inscribed square, in the directions shown in the diagram. Two equal forces also act at points A and B, along the line of the diagonals of the square. The forces acting along the sides of the square

are each of 1 kg, and those acting along the diagonals each equal $\sqrt{2}$ kg. Find the resultant of all the forces and its point of application. How will the ring move under the action of the forces given?

29. A scale-pan is attached to a spiral spring, whose extension is subject to Hooke's law; in the scale-pan is a weight (Fig. 12). With what force should the scale-pan be pulled downwards so that when it is released there should be a moment at which the weight ceases to exert pressure on the scale-pan.

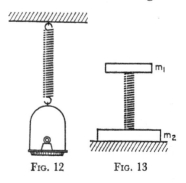

FIG. 12   FIG. 13

30. Two laminas of mass m_1 and m_2 are joined by a spring (Fig. 13). With what force should the upper lamina be pressed downwards so that when the force is removed the upper lamina should spring back and raise the lower lamina a little too? The coefficient of elasticity of the spring is k. Assume that Hooke's law is applicable throughout. Ignore the mass of the spring.

31. A cyclist moves with uniform velocity down an inclined plane. What is the size and direction of the plane's reaction?

32. A plank inclined at an angle of α to the horizontal lies on two supports A and B (Fig. 14), over which it can slip without friction under the action of its own weight Mg. With what acceleration and in what direction should a man of mass m move along the plank so that it should not slip?

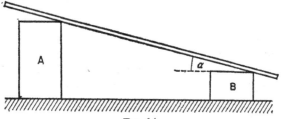

FIG. 14

33. A fly is sitting at the bottom of a test-tube. The test-tube falls freely, maintaining its vertical position (Fig. 15). How will the duration of the test-tube's fall be affected if the fly, during the test-tube's fall, flies up from the bottom of the test-tube to the top?

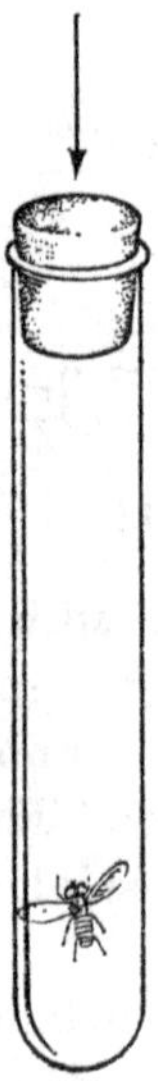

34. A bird is enclosed in a box standing on one pan of a pair of scales. While the bird is sitting on the bottom of the box, the scales are balanced by weights in the other scale-pan. What will happen to the scales if the bird takes off and hovers inside the box?

35. A balloon descends with constant velocity v. What amount of ballast must be jettisoned from the balloon so that it should rise with the same velocity v? The air-resistance is proportional to the velocity. The weight and carrying capacity of the balloon are known.

36. A bullet travels vertically upwards, reaches its highest point and falls back vertically downwards. At what points of its trajectory does the bullet's acceleration have its maximum and its minimum value? Take into account air-resistance, which increases in proportion to the increase of the bullet's velocity.

FIG. 15

37. A spring is put into a large tube and occupies the tube's full length when not subject to outside forces. A sphere is placed on top of the spring and compresses it to approximately half its previous length (Fig. 16). Then the tube begins to fall in an inclined position. What will happen to the sphere?

38. A balance is mounted on a stationary trolley, with a weight suspended from one end, while the other end is linked to the floor of the trolley by a spring (Fig. 17). If the trolley be accelerated in a horizontal direction by a constant force, the weight will be inclined at an angle in the direction opposite to the line of acceleration. Will this alter the tension of the spring?

39. A piston is fixed in the cylindrical part of a vessel containing compressed air. The volume of the cylindrical part is

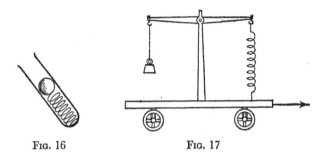

FIG. 16   FIG. 17

small by comparison with that of the whole vessel (Fig. 18). If the piston be released from the forces which hold it in place, it will be pushed downwards out of the vessel (there is no friction between the sides of the cylinder and the piston).

How will the time taken for the piston to move down the cylindrical part be affected if: (1) a small sphere be placed on top of the piston? (2) the weight of the piston be increased by an amount equal to the weight of the small sphere?

40. Two boys A and B attach a dynamometer by a ring to a nail driven into a wall and fasten a cord to the dynamometer's 'hook; they then take turns to pull the cord to see which of them is stronger. When A pulls, the dynamometer registers 42 kg, and when B pulls it registers 35 kg. What will it register if the boys take it down

FIG. 18

from the nail and take hold, one of the cord and the other of the hook, and then pull in opposite directions (Fig. 19)? (In neither of the cases given do their feet slip on the ground.)

41. In the film 'Brave People' the hero of the film jumps from a train moving along a level track on to the buffer-mounting and uncouples the last two carriages. In what cases is this possible?

FIG. 19

42. Two weights of mass m_1 and m_2 are joined by a non-elastic cord passing over a fixed pulley (Fig. 20). Find the acceleration of the loads, the tension in the cord and the force exerted on the pulley's axle. Ignore the mass of the pulley.

43. Through the middle of a rod of length $2l = 2$ m, passes a horizontal axle O, about which the rod can rotate. To the ends of the rod are fixed loads M_1 and M_2 of mass 1 and 7 kg re-

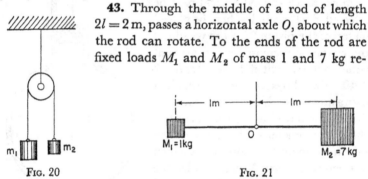

FIG. 20   FIG. 21

spectively (Fig. 21). The rod is brought into a horizontal position and then smoothly released. What force will it exert on the axle at the instant after release? Neglect the mass of the rod and the friction in the axle.

44. A chain is lying on an absolutely smooth table, half of it hanging over the edge of the table (Fig. 22a). How will the time it takes to slip off the table be affected if two weights of equal mass be attached, one to each end (Fig. 22b)?

45. A rope passes over a weightless pulley A, with a load M_1 attached to one end and to the other a weightless pulley B, carrying loads M_2 and M_3 on the ends of *its* rope. The whole system is hung, by pulley A,

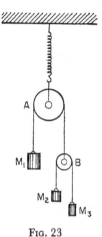

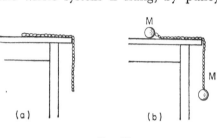

Fig. 22

Fig. 23

from a spring balance (Fig. 23). Find the acceleration a of load M_1 and the reading T of the spring balance, taking $M_2 \neq M_3$, $M_1 > M_2 + M_3$.

46. A homogeneous chain of length l and mass m hangs partly from a table and is held in equilibrium by the force of friction. Find the coefficient of static friction if it be known that the greatest length of the chain that can be hanging from the table without the whole chain slipping is l_1.

47. If a locomotive cannot move a heavy train from rest, the driver acts as follows: he puts the locomotive into reverse and then, having pushed the train back a little, switches into forward gear. Explain why this procedure allows the train to be moved forward.

48. According to Newton's law only an outside force impressed by another body can alter the state of motion of a given body. Then what outside force brings a car or any other self-moving vehicle to a stop under braking?

49. A long-handled broom lies horizontally on the forefingers of a pair of hands held wide apart (Fig. 24). What will happen if the left hand remains still and we move the right hand towards the left, keeping it constantly at the same level? What will happen if the right hand remains still and we move the left hand towards it? What will happen if we move both hands towards one another at the same time?

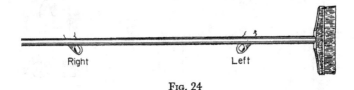

Right   Left

Fig. 24

50. If a fast-moving car brakes sharply, its forward part dips. Why does this happen?

51. A small body slips, subject to the force of friction, from point A to point B along two curved surfaces of equal radius,

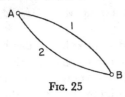

Fig. 25

first along route 1, then along route 2 (Fig. 25). The friction does not depend on the speed and the coefficient of friction on both routes is the same. In which case will the body's velocity at B be greater?

52. Two identical weightless pulleys are mounted with parallel axes at the same height. A non-elastic weightless rope is passed through both pulleys with identical weights on the ends (Fig. 26).

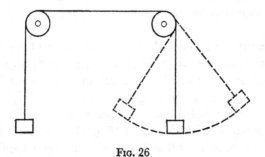

Fig. 26

The system is in equilibrium. One of the weights is pulled to one side and released. Is equilibrium broken?

53. The driver of a car travelling at velocity v suddenly sees a broad wall in front of him at distance a. Is it better for him to brake or to turn sharply?

III. Statics

54. Can a man, standing against a wall so that his right shoulder and right leg are in contact with the wall (Fig. 27), raise his left leg and in so doing not lose equilibrium?

Fɪɢ. 27

55. In what cases could the heroes of Krylov's famous fable, the swan, the pike and the crab, not have moved the cart in fact, assuming that they are all of equal strength and that there is no friction between cart and ground?*

56. To remove a trolleybus's arm from the wire, the driver first of all tugs the rope attached to the ring which slides up and down the arm as far back as possible up the arm. Why?

57. A lamp hangs from a bracket whose three arms each have one end fixed in the wall, the other ends meeting at a point. The two upper arms form an isosceles triangle with an angle of

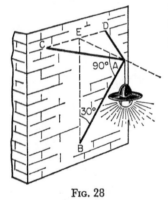

Fɪɢ. 28

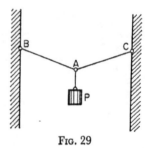

Fɪɢ. 29

* In this fable, a swan, a pike and a crab pull at a cart in three opposing directions. The cart does not move.

60° between the arms. The plane of this triangle is at right angles
to the third arm, which makes an angle of 30° with the wall.
The bulb and shade weigh 1 kg. Find the stresses in the arms
(Fig. 28).

58. A load is attached to two strings *AB* and *AC* of equal length
and suspended from them (Fig. 29). In what case will the strings
break most easily, when they hang down almost vertically, or
when they are stretched almost horizontally? Neglect the weight
of the strings.

59. The following trick is sometimes used to move a car which
is stuck : a rope is attached to the car and to a tree, the rope being
stretched as taut as possible. Then if a man pulls on the rope in a
direction almost at right angles to it, he will easily move the car.
Why is this possible?

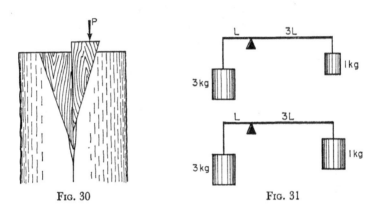

FIG. 30   FIG. 31

60. Using the parallelogram of forces, show that one wedge
can be driven out by another (Fig. 30).

61. Two bars are in equilibrium (Fig. 31). On the first are
balanced two different weights made of the same material, on the
other two weights of different mass but equal volume. Will their
equilibrium be disturbed if the systems be lowered into water?

62. Weights in equilibrium, together with the beam from which

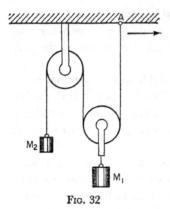

Fig. 32

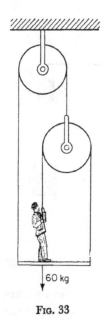

60 kg

Fig. 33

they hang, are placed along a magnetic meridian. Will equilibrium be preserved if the beam is magnetized along its length?

63. When one of the loads is very slightly greater than the other, the scale-pan of a beam-balance on the overloaded side dips slightly and remains in equilibrium in that position. Why is equilibrium reached even though the weights of the loads are different?

64. In a system consisting of one fixed and one movable pulley, the loads are in equilibrium when the ropes are parallel. What will happen if the point of attachment of the rope at A be moved to the right (Fig. 32)? Neglect the weight of the pulleys.

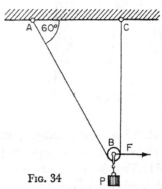

Fig. 34

65. With what force must a man pull on the rope to hold the board in position if the man weighs 60 kg (Fig. 33)? Neglect the weight of board, rope and pulley.

66. With what horizontal force F must a small pulley be drawn aside (B in Fig. 33), so that section BC of the rope is to be vertical, if a load P be hung from the pulley and the section AB of the rope is to form an angle of 60° with the horizontal AC?

67. A thin homogeneous lamina is in the form of a circle of radius R; in it a circular hole is cut of exactly half the radius of the lamina and touching the lamina's circumference (Fig. 35). Where is the centre of gravity?

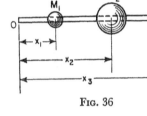

FIG. 35   FIG. 36

68. A thin weightless bar passes through the centres of three spheres of different masses, M_1, M_2 and M_3. The centre of the first sphere is at a distance of x_1 from the end of the bar, the centre of the second at a distance of x_2 and the centre of the third at a distance of x_3 (Fig. 36). At what distance from the same end of the bar is the centre of gravity of all three spheres?

69. A cylinder of weight P lies on a plane inclined at an angle of 30° to the horizontal. The cylinder is held in place by the help of a rope passing round it (Fig. 37) of which one end is fastened to the plane while the other is stretched vertically upwards with force Q. What is the value of force Q?

70. A wheel of radius R and mass m stands in front of a step of height h (Fig. 38). What is the least horizontal force F which must be applied to the axle of the wheel (O) to allow it to rise on to the step? Ignore the force of friction.

71. In both cases shown in Fig. 39 mass m_1 is so chosen that mass m_2, which rests on a smooth surface, is in equilibrium. In which case is equilibrium stable and in which case is it unstable? For simplification we shall suppose that the pulley is sufficiently far

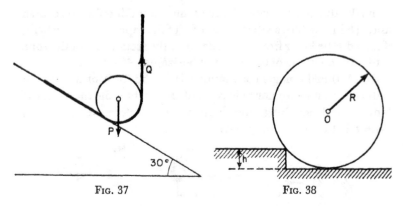

Fig. 37   Fig. 38

away, and that therefore the line of the rope from mass m_2 is parallel to the line of the tangent to the surface.

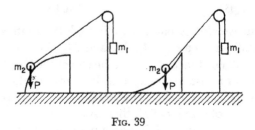

Fig. 39

72. A sideboard has a sliding board in it for cutting bread on. For convenience in pulling out the board there are two handles

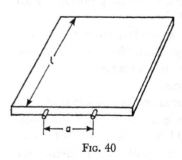

Fig. 40

in the front edge distance a apart and placed symmetrically in relation to the centre (Fig. 40). The length of the leaf is L. What is the least coefficient of friction k between the side of the leaf and the wall of the sideboard which will prevent the leaf being pulled out by one handle only, no matter how much force is applied?

IV. Work, Power, Energy. The Law of Conservation of Momentum. The Law of Conservation of Energy

73. A cylinder and a cube of the same material, the same height and the same weight stand upright on a horizontal plane. Which of the two bodies is it harder to overturn?

74. Calculate the minimum amount of work necessary to overturn a crate of weight 1 ton, first about edge AB, then about edge $A'B'$. The dimensions of the crate are given in Fig. 41.

75. Will the work and power expended by the motor of a moving-staircase change if a passenger standing on it as it moves upwards himself walks up the staircase at a constant speed?

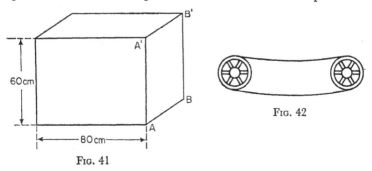

FIG. 41

FIG. 42

76. Two pulleys on the same horizontal level are connected by a belt (Fig. 42). The left-hand pulley is the driving one. Is it possible to transmit more power when the pulleys are revolving in a clockwise or a counter-clockwise direction?

77. The principle of the automatic weapon is based on the utilization of the phenomenon of recoil which takes place on

firing: the breech-block moves backwards after firing and compresses the spring, which in turn brings the reloading mechanism
into action. Find what must be the velocity of the bullet for the
breech-block to move back a distance *a*, if the mass of the bullet
be *m*, the mass of the breech-block *M* and the coefficient of
elasticity of the spring *k*. Neglect the mass of the charge.

78. Two identical springs, one of steel, the other of copper,
are stretched by an identical amount. On which operation must
more work be expended?

79. Two identical springs, one of steel, the other of copper, are
stretched with identical forces. On which operation must more
work be expended?

80. A load of mass *m* is suspended from a thread of length *l*.
Find the least height to which the load must be raised so that it
should break the thread in falling, assuming that the least load
which would break the thread when simply suspended from it is
M and that this load would stretch the thread by 1 per cent of
its length at the moment of breaking. Assume that Hooke's law
applies to the thread right up to breaking-point.

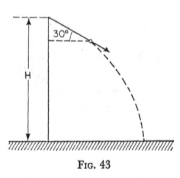

FIG. 43

81. Starting from a height *H*,
a ball slips without friction,
down a smooth plane inclined at
an angle of 30° to the horizontal
(Fig. 43). The length of the plane
is *H*/3. The ball then falls on to
a horizontal surface with an impact that may be taken as perfectly elastic. How high does the
ball rise after striking the horizontal plane?

82. A bullet of mass *m* hits a wooden block of mass *M*, which is
suspended from a thread of length *l* (a ballistic pendulum), and is
embedded in it. Find through what angle the block will swing if
the bullet's velocity is *v* (Fig. 44).

83. A test-tube of mass *M*, closed with a cork of mass *m*, con-

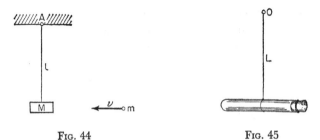

FIG. 44   FIG. 45

tains a drop of ether. When the test-tube is heated, the cork flies
out under the pressure of the ether-gases. The test-tube is suspended by a weightless, rigid bar of length L (Fig. 45). What is
the least initial velocity which will cause the test-tube to describe a full circle about the pivot O?

84. Two light trolleys of weights m_1 and m_2 ($= 3m_1$) are connected by a spring (Fig. 46). The spring is compressed and bound
by a cord. The cord is burnt through, the spring expands and the
trolleys move off in opposite directions. Find (a) the relationship
between the velocities v_1 and v_2 of the trolleys; (b) the relationship
between the periods t_1 and
t_2 during which the trolleys
move; (c) the relationship
between the distances s_1
and s_2 travelled by the
trolleys. Assume that the
coefficient of friction is the
same for both trolleys.

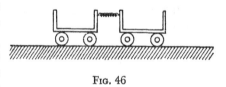

FIG. 46

85. A shell fired at a certain angle to the horizontal, bursts into
two fragments of equal mass at the top of its parabolic trajectory.
One fragment returns to the point of firing as a result of the
explosion, following its original trajectory. Where will the other
fragment fall? Will both fragments hit the ground at the same
moment? Neglect air-resistance.

86. A stationary ball is truck obliquely (i.e. not along the line
joining their centres) by another ball of the same mass. At what

angle will the subsequent paths of the balls diverge, assuming that
the balls are absolutely elastic and absolutely smooth?

87. Several identical balls of steel or bone are suspended by
threads from a board (Fig. 47). In their initial position, the balls
touch one another and the threads are all parallel. If the end ball
on the right be drawn aside and
released, after its impact on the
line of stationary balls it remains
at rest, but the end ball at the
left is knocked sideways. But if
the end two balls, instead of one,
at the right are drawn aside and
released simultaneously, then
two balls will be knocked aside
at the left. How can this experiment be explained?

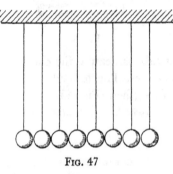

FIG. 47

88. Suppose that the spheres of problem 68 move uniformly
along parallel trajectories with velocities of v_1, v_2 and v_3 respectively. With what velocity will the centre of gravity of these
spheres move?

89. Making use of the classic formulas for the velocities of two
spheres after impact:

$$v'_1 = \frac{(m_1 - m_2)v_1 + 2m_2 v_2}{m_1 + m_2}, \qquad v'_2 = \frac{(m_2 - m_1)v_2 + 2m_1 v_1}{m_1 + m_2},$$

where v_1, v_2 are the velocities of the spheres before impact, v'_1,
v'_2 their velocities after impact—show that the velocity of the
centre of gravity of two spheres after impact (regardless of the
nature of the impact) equals the velocity of the spheres' centre of
gravity before impact.

90. A rope is passed through a pulley which is suspended
sufficiently high. Two monkeys of equal weight climb the rope
from opposite ends, one of them climbing more quickly than the
other, relative to the rope. Which of the monkeys will reach the

top first? Assume that the pulley is weightless and that the rope is both weightless and inextensible.

91. A boat is at rest in stagnant water. A man in the boat walks from the bow to the stern. What distance will the boat move if the man's mass, m, = 60 kg, the boat's mass, M, = 120 kg, and the length of the boat, L, = 3 m? Neglect the resistance of the water.

92. Two identical guns are mounted on a railway-truck which is free to move along the rails, and point in opposite directions (Fig. 48). The sights are set so that, if the guns are fired simul-

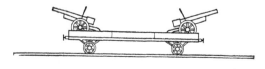

Fig. 48

taneously, each will score a hit on its target. Will the shells still hit their targets if one of the guns fires slightly before the other? What will happen to the truck after the second shot? Neglect friction in the wheels of the truck.

93. Three balls of equal mass are suspended from a thread and two springs of the same elasticity so that the distances between the first and second ball and the second and third are the same (Fig. 49). Thus the centre of gravity of the whole system coincides with the centre of the second ball. If the thread supporting the top ball be cut, the system will fall and the acceleration of the system's centre of gravity will be $\dfrac{mg + mg + mg}{3m} = g$ (according to Newton's second law, the acceleration of the centre of gravity of a system equals the sum of the forces acting on the system from outside divided by the system's total mass). But spring I will pull the

Fig. 49

second ball upwards with greater force than spring *II* will pull it downwards (the force of spring *I* at the initial moment $f_{10} = 2mg$, while the force of spring *II* at the initial moment $f_{20} = mg$) and therefore the centre of the second ball will have, at the initial moment, an acceleration of less than *g*. And yet the centre of gravity of the whole system must move with an acceleration of *g* the whole time. Explain the contradiction.

94. A pocket-watch is placed on a horizontal stand which can rotate freely about its vertical axis (Fig. 50). How will this affect the watch's working? Neglect the friction of the stand about its axis.

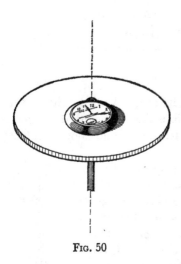

FIG. 50

95. A small ball is thrown vertically up with a given initial velocity. At the moment when it reaches the highest point of its flight, another similar ball is thrown upwards with the same velocity and in the same vertical line. The two balls collide at a given point and at this moment a third similar ball is thrown upwards from the same point, with the same velocity and in the same vertical line. How long after the throwing of the third ball will all three balls fall, one after the other? The impact of the balls on collision is perfectly elastic.

96. By the principle of Newtonian relativity, two systems of co-ordinates moving uniformly and in a straight line relative to each other are equivalent, i.e. the physical laws which hold good in one system, also hold good in the other. Let system *II* be moving, relative to system *I*, uniformly and in a straight line with velocity *v*. Body *A* is moving in the same direction with velocity v_1 relative to system *I* (and therefore with velocity $v_1 - v$ relative to system *II*). A constant force F acts on the body A for a certain period t, in the same line as velocities v and v', and changes the body's velocity relative to system I from v_1 to v_2. The body's change in kinetic energy will be:

in system *I*

$$\frac{m}{2}\,(v_2^2 - v_1^2),$$

and in system *II*

$$\frac{m}{2}\,(v_2-v)^2 - (v_1-v)^2 = \frac{m}{2}\,(v_2^2-v_1^2) - mv(v_2-v_1),$$

i.e. less. The change in kinetic energy is therefore different in different systems of co-ordinates. How can this be reconciled with Newton's principle of relativity?

V. The Dynamics of Motion in a Circle

97. To reduce losses due to friction when a shaft rotates in its bearing, it is proposed to shape the end of the shaft towards the form of a cone (Fig. 51). Then the magnitude of the force of friction is obviously unchanged inasmuch as the reduction in the area of the surfaces in contact is accompanied by an increase in the specific pressure between them and a corresponding increase in the force of friction per unit area of contact. However this shaping of the shaft *can* reduce losses due to friction, supposing that there is no friction against the side walls of the bearing. Why is this so?

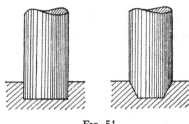

Fig. 51

98. A turning moment of 100 kgm is applied to a shaft which passes through a wheel of radius $R = 50$ cm. With what force must brakes be applied to the wheel (Fig. 52) so that the wheel should not rotate? The coefficient of friction equals 0·25.

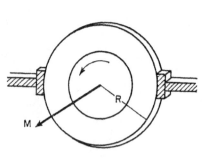

Fig. 52

Fig. 53

99. A rope is passed through a fixed pulley whose moment of inertia is l and from each end of the rope hang loads of unequal mass, m_1 and m_2 (Fig. 53). What must be the tension in the rope on either side of the pulley?

100. The turning of a car must be produced by an external force acting at an angle to the line of motion of the car. This external force can only be the force of friction between tyres and ground. Then why, when the front wheels of the car are turned, does the direction of the force of friction change in such a way that the whole car is caused to turn?

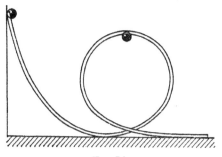

FIG. 54

101. A body slips down a chute which is in the form of a loop as in Fig. 54. It starts from the lowest admissible height, and follows the loop without falling. In doing this, it does not exert any pressure on the loop at the highest point of its path. On what body is centrifugal force acting at this moment.

102. A plumb-line is set up on a rotating disk and makes an angle of α with the vertical, as in Fig. 55. The distance r from the point of suspension

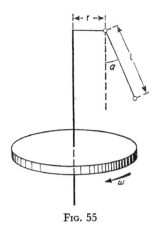

FIG. 55

to the axis of rotation is known, and so is the length *l* of the thread. Find the angular velocity of rotation.

103. Why does a coin rolling along a plane in a vertical position (i.e. without tilt), move along a straight line, while one that is tilted follows a curved path?

104. A skater described a circle on ice by leaning towards the centre of the circle. What is the source of the centripetal force necessary to motion in a circle?

105. A pilot of weight 70 kg loops a loop of radius 100 m in an aeroplane flying at 180 km/hr. What is the source of the centripetal force acting on the pilot at the highest and lowest points of the loop? On what is the centrifugal force acting at these points? And with what force is the pilot pressed to his seat at the highest and lowest points of the loop?

106. A sphere of mass *m*, suspended from a thread of length *l*, is drawn sideways from its position of equilibrium so that it is raised through height *h* (Fig. 56). Then the sphere is released. To what height will it rise if a bar *A* be placed in the path of the thread perpendicular to the plane of the sketch Galileo's experiment)?

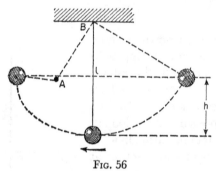

FIG. 56

107. A small ball is suspended from point *A* by a thread of length *l*. A nail is driven into the wall at a distance of *l*/2 below *A*, at *O*. The ball is drawn aside so that the thread takes up a horizontal position (Fig. 57). At what point in the ball's trajectory will the tension in the thread disappear? How much farther will the ball move. What will be the highest point to which it will rise? At what point will the ball pass through the vertical line passing through the point of suspension?

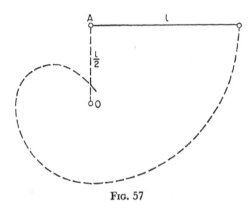

Fig. 57

108. Why is it that when a billiards-ball is struck towards the underside by the cue, it moves slowly, and when it is struck towards the upper side it at first moves with greater speed?

109. The circumference velocity of points on the earth's equator is about 460 m/sec. What would happen to a bullet fired from a gun parallel to the earth's surface towards the west with this same velocity if there were no atmosphere?

110. Given two cylinders of identical dimensions and material. One of them is solid, the other is composed of two cylinders fitting together almost without clearance and without friction between them (Fig. 58). Which of the cylinders will roll (without slipping) faster down the same inclined plane? What position does the inside cylinder adopt while the rolling takes place, if it does not fit exactly inside the containing cylinder?

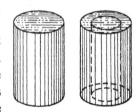

Fig. 58

VI. The Universal Theory of Gravitation

111. A spherical hollow is made in a lead sphere of radius R, such that its surface touches the outside surface of the lead sphere and passes through its centre. The mass of the sphere previous to hollowing was M. With what force (according to the universal law of gravity) will the lead sphere attract a small sphere of mass m, which lies at a distance d from the centre of the lead sphere on the straight line connecting the centres of the spheres and the side of the hollow (Fig. 59)?

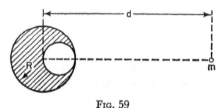

FIG. 59

112. The sun attracts objects on the surface of the earth with a greater force than the moon. Yet the phenomenon of ebb and flow is caused chiefly by the agency of the moon, not of the sun. Why?

113. The force of the sun's attraction acts on all bodies on the earth. At night (when the sun is 'beneath our feet') this force is combined with the force of the earth's attraction, by day (when the sun is 'above our heads') it works against the earth's attraction. Therefore, at night all objects should be heavier than they are by day. Is this true?

114. A body is raised, with the help of a rocket, to a height of 500 km. (1) What is the acceleration due to the force of gravity at this point? (2) With what velocity should this body be projected in a direction perpendicular to the radius of the earth, so that it should describe a circle about the earth? (3) What will be the period of revolution of the body about the earth in (2)? Take the radius of the earth as being 6500 km. Neglect the resistance of the atmosphere.

In this problem, the body will be subject to conditions almost identical to those which obtained for the first Soviet artificial earth-satellite. The acceleration of the force of gravity at the earth's surface should be taken as 98 cm/sec.

VII. Oscillation, Waves, Sound

115. A load of mass m lies on a perfectly smooth plane, being pulled in opposite directions by springs 1 and 2, whose coefficients of elasticity are k_1 and k_2 respectively (Fig. 60). If the load be forced out of its state of equilibrium (by being drawn aside), it will begin to oscillate with period T. Will the period of oscillation

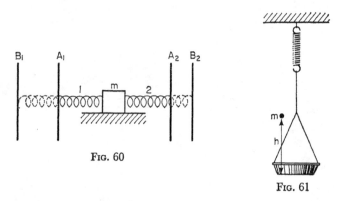

Fig. 60

Fig. 61

be altered if the same springs be fastened not at points A_1 and A_2, but at B_1 and B_2? Assume that the springs are subject to Hooke's law for all strains.

116. A load of mass m falls from height h on to a scale-pan suspended from a spring whose coefficient of elasticity is k; the load remains on the pan, i.e. its impact on the bottom of the scale-pan may be considered perfectly inelastic (Fig. 61). The pan begins to oscillate. Find the amplitude of the scale-pan's oscillation. (For the purposes of simplification consider first the case when the pan's weight may be neglected.)

117. How will the period of oscillation of a pendulum consisting of a vessel suspended on the end of a long thread alter, if the vessel is filled with water which gradually flows out of a hole in the bottom (Fig. 62)?

118. Two identical pendulums are connected by a weightless spring (Fig. 63). In one case both pendulums oscillate in such a way that at any moment they are inclined to the same side at the same angle. In the other case they oscillate in such a way that they are inclined at the same angle, but to opposite sides. In which case will the period of oscillation be less?

119. Two identical pendulums connected by a light spring (Fig. 63) are drawn to one side through an identical angle in the same plane as the figure and, as a result, they oscillate in the same plane as the figure. Will the period of the oscillations be increased or decreased if one of the pendulums be removed and the spring be secured at its mid-point?

Fig. 62

120. A stand makes simple harmonic oscillations in a vertical line (Fig. 64), the amplitude of the oscillations being $A = 0.5$ m. What must be the least period of these oscillations if an object lying on top of the stand is not to be separated from it.

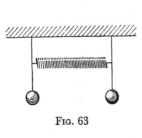

Fig. 63

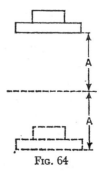

Fig. 64

121. A stand makes simple harmonic oscillations in a horizontal line with a period of $T = 5$ sec. A body on the stand begins to slip across it when the amplitude of the oscillations reaches a value of $A = 0.6$ m. What is the coefficient of friction between the body and the stand?

122. How will the period of oscillation of a pendulum be altered if its point of suspension be moved (a) vertically upwards with acceleration a; (b) vertically downwards with acceleration $a < g$; and (c) horizontally with acceleration a?

123. A swinging pendulum is suspended from a massive board. At a given moment, the board starts to fall freely. How will the pendulum then continue to move relative to the board? Distinguish between two cases: the board begins to fall (1) when the bob is in one of its two rest-positions, and (2) when the bob is at some intermediate position. Neglect the force of friction.

124. Find the period of oscillation of a ball slipping down and then up two inclined planes (Fig. 65). Leave friction and loss of energy upon impact out of account.

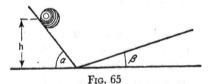

FIG. 65

125. The founder of Russian seismology B. B. Golitsyn perfected a so-called horizontal pendulum which registered earthquakes taking place at great distances from the location of the pendulum. This pendulum has an axis of oscillation which forms a small angle with the vertical α. The pendulum can be represented diagrammatically as an equilateral triangle with one side as the pendulum's axis and a weight attached to the opposite vertex (Fig. 66). The sides of the equilateral triangle can be considered weightless. The length of side is L. Find the period of small oscillations of the pendulum.

126. A pendulum consists of a metal ball suspended by a long silk thread. How will the period of the pendulum's oscillations be altered if the ball be charged negatively and another, positive, charge be placed (a) below the pendulum in a vertical line with

the point of suspension (Fig. 67a); (b) at the point
of suspension (Fig. 67b); (c) to one side, on a level
with the ball's lowest position, so that the ball can
not touch it (Fig. 67c).

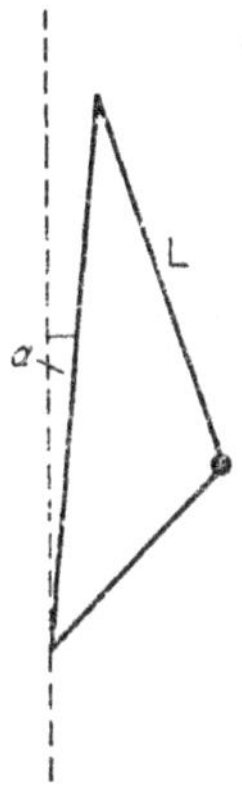

127. A radio station located at point A sends out
a time-check, which is received by two sets at points
B and C (Fig. 68). A listener at B hears the signal
from his own set and 1 sec later hears the sound of
the same signal being received at C by the set there,
which has a powerful loudspeaker. What is the
distance between points B and C?

128. It is required to tune a string to unison
with a tuning-fork and both are accordingly made
to sound at the same time. Beats occur. After a
small weight has been attached to one prong of the tuning-fork,
the frequency of the beats is reduced. What must be done to the

FIG. 66

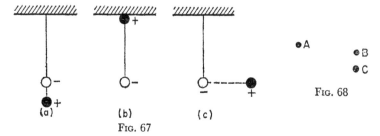

(a)   (b)   (c)
FIG. 67

•A
•B
•C
FIG. 68

string (tauten or slacken it) so that it should be tuned
to unison with the tuning-fork?

129. Is it always true that 'an echo doesn't lie'—i.e.
that a reflected sound has the same pitch as the original
direct sound?

130. Why is a tuning-fork made with two prongs
(Fig. 69)? Would a tuning-fork be of any use for its
normal purpose if one of the prongs were sawn off?

131. One of O'Henry's heroes kicked a pig so hard
that it took off and 'overtook the sound of its own

FIG. 69

squeal'. With what force should his hero have had to kick the pig in reality for this to have happened? Take the mass of the pig as being 5 kg, and the duration of the impact as 0·01 sec.

132. Where ought a man to hear a louder sound, at the anti-node or at the node of displacement of a stationary wave?

133. Why does the sound in a hall filled with people sound deader than in the same hall empty?

VIII. The Mechanics of Liquids and Gases

134. Does the law of communicating vessels (that a homogeneous liquid contained in communicating vessels will be at the same level in each) hold good if there is a float on the surface of the liquid in one of the vessels? Neglect capillarity.

135. A sufficiently long tube, open at the lower end and with a tight-fitting piston which can move inside the tube without friction, is under water, suspended by a string (Fig. 70). The upper part of the tube above the piston is empty. How is the tension in the string affected by the depth of the tube's immersion?

136. A bucket is being drawn up a well. What force must be applied to raise the bucket: (1) while it is still under water, and (2) when it has been drawn out of the water? The bucket weighs P, is made of material of specific gravity d and has a capacity V. Neglect the resistance of the water to the bucket's motion.

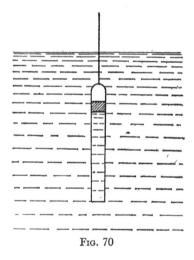

Fig. 70

137. In order to raise the level of a liquid in a vessel through a height h, by means of a pump, a certain amount of work must be done. Will the amount of work required for this be altered if a body is floating on the surface of the liquid?

138. Will the draught of a steamer be altered as a result of a change in the force of gravity's acceleration, if it changes latitude by sailing from northern to equatorial waters?

139. How does the lifting power of an aerostat depend on the temperature at which the lift takes place?

140. When people are explaining an experiment to weigh air, they sometimes say that first of all a flask is weighed with air in it and then, when the air has been pumped out, the flask alone is weighed. The difference in the readings of the first and second weighing is the weight of the air in the flask.

Is this a correct explanation of the experiment of weighing air?

141. Is it possible to measure the density of air by weighing a soft, airtight sac at first when it is empty (compressed), and then when it is filled with air? The capacity of the sac when filled is known.

142. The tube of a mercury barometer is hung from a hook on one of the pans of a pair of scales. Find what weights should be put in the other pan if the scales are to be in equilibrium (Fig. 71).

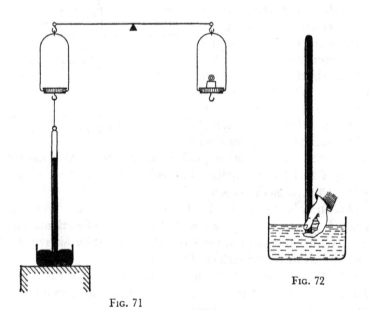

Fig. 71

Fig. 72

143. Will Torricelli's experiment work if a barometric tube filled with mercury be placed with its open end not in a vessel containing mercury, but in vessel containing water (Fig. 72)?

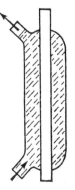

144. Why is it not possible to extinguish burning kerosene, which is lighter than water, by pouring water on it?

145. In many distilling apparatuses the tube in which the steam is condensed is surrounded with another tube ('jacket'), through which cold water circulates (Fig. 73). The water is passed through the jacket from bottom to top. Why not the other way about?

FIG. 73

146. It is discovered that the level of water in both limbs of a U-tube with both ends sealed is the same both when the tube is vertical and also when it is tilted in a vertical plane (Fig. 74). In what conditions can this occur?

147. The shorter limb of a curved tube is covered with a very thin, soft and impermeable membrane. The tube is filled with hydrogen and set with its open end downwards (Fig. 75). What position will the membrane adopt?

148. Vessels A and B contain carbon dioxide (CO_2) and hydrogen (H_2). Pressure-gauges M_1 and M_2 register equal

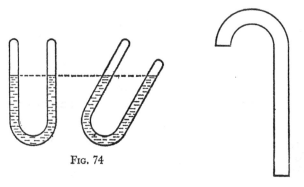

FIG. 74

FIG. 75

pressures. In which direction will gas flow if the valve K is opened (Fig. 76)? What will happen if the same experiment is carried out with the vessels turned so that the pressure-gauges are underneath?

149. Why are gas-towers not built on the lines of water-towers?

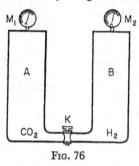

FIG. 76

150. In some cases when a deep well has been bored in the earth, compressed air is fed into the well through a tube lowered into it. Will the excess pressure (i.e. the difference between the pressure inside the tube and outside it) be the same at the bottom of the tube as at the top or not? Neglect the loss of pressure in the tube due to the flow of the air along it.

151. In some houses gas burns better in the basement than on the upper floors. Why is this?

152. Two balloon envelopes of identical weight, one made of thin rubber and the other of rubberized fabric, are filled with identical quantities of hydrogen and at ground level occupy an equal volume. Which of the balloons will rise the higher, if the hydrogen is unable to escape?

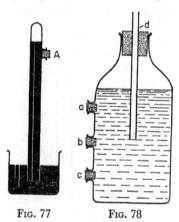

FIG. 77   FIG. 78

153. A bubble-level consists of a slightly bent tube filled with liquid so that a very small space is left – a bubble in which is the liquid's vapour. When is the bubble larger, in winter or in summer?

154. In the barometer shown in Fig. 77, the opening at A is closed with a cork and the barometer is filled with mercury. What will happen if the cork is removed from opening A?

155. There are three small holes in the side of a Mariotte's vessel (Fig. 78), closed by corks, *a*, *b* and *c*. A tube *d*, open at both ends, passes through the cork in the neck of the vessel. The level of the water in the Mariotte's vessel is above hole *a*. The level of the water in the tube *d* is at the very bottom of the tube, which is on a level with hole *b*. What will happen if one of the holes *a*, *b* or *c* is unstopped?

156. To measure the volume of a powder which does not react with air and does not absorb air, an apparatus is used which is called a volumometer. This apparatus is illustrated in Fig. 79. A flask *A* is fitted, after grinding to give an exact fit, to a bent glass tube which has a tap at *K* and ends in a bulb *B*, which turns into a straight tube. This straight tube is connected by a length of rubber tubing to another straight glass tube which can be moved vertically up and down a scale. Lines are marked at the upper and lower limits of the bulb, the volume between them being measured exactly and found to equal *V*. Both the straight glass tubes and the rubber tubing contain mercury, thus making a pressure-gauge.

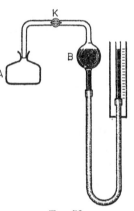

FIG. 79.

The volume of the powder is measured in the following way. Atmospheric pressure *H* is marked on the pressure-gauge. With tap *K* open the level of the mercury is brought to the upper mark on the bulb *B* and the tap is then closed.

Then the straight tube of the pressure-gauge is lowered down the scale until the level of the mercury in the bulb reaches the lower mark; the difference in the levels of the mercury in the straight tubes is then noted (*h*).

After this the apparatus is restored to its original position and the tap *K* is opened.

The flask A is then removed and the powder of which we wish to find the volume is poured into it. The flask is again connected to the tube and when the mercury in both tubes is on a level with the upper line on the bulb, the tap K is closed. The pressure-gauge's straight tube is lowered once more so that the level of mercury in the bulb B reaches the lower line and h', the new difference in levels in the straight tubes, is measured.

How can we find the volume of the powder, knowing the values of V, H, h and h'?

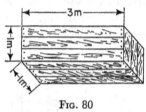

FIG. 80

157. An empty box, open on the underside, is dipped into water in a vertical position in such a way that the lid of the box is at a depth of 18·6 m. The box's dimensions are given in Fig. 80. Find the upthrust acting on the box.

158. When determining the specific gravity of a solid body with the help of hydrostatic scales, the body is first weighed in air (on ordinary scales) and then when immersed in water. How must this method be altered if the body's specific gravity is less than unity?

159. Two holes are made in the side of a cylindrical bucket, symmetrically placed on opposite sides. The holes are closed with corks and the bucket is filled with water. If the corks be taken out the water will stream out of the holes. What will happen if the bucket falls freely?

160. An upturned test-tube is fixed rigidly over a vessel filled with water (Fig. 81). How will the level of water in the test-tube be altered if the whole system begins to fall freely?

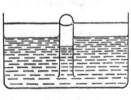

FIG. 81

161. A submarine which comes to rest on a sea-bed of clay or sand can sometimes not raise itself again. How is this phenomenon of 'sucking down' to be explained?

162. One limb of a U-shaped open-ended mercury manometer is connected to a flask with water in it, the air having been pumped out (Fig. 82). What will the manometer register if the flask be lowered into a vessel containing boiling water? Will the manometer's reading depend on height above sea-level?

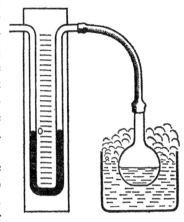

163. Water at 0°C can be raised by a suction pump through a height of 10 m. Through what height, greater or less, can the pump raise hot water at 90°C?

Fig. 82

164. A piece of ice is floating in a beaker, filled to the brim with water (Fig. 83). Will the water overflow the rim of the beaker when the ice melts? What will happen if the beaker is filled not with water, but with (1) a liquid denser than water (2) a liquid less dense than water?

165. A piece of ice is floating in a vessel containing water, and inside the ice is a piece of lead (Fig. 84). Will the level of the water in the vessel be altered when the ice melts? What will happen if there is not lead inside the ice, but bubbles of air?

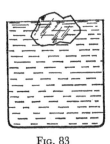

Fig. 83

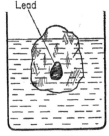

Fig. 84

166. On one scale-pan of a pair of scales is a vessel containing water, and on the other is a stand, from the arm of which a body is suspended by a weightless thread (Fig. 85). While the body is not in the water, the scales are in equilibrium. Then the thread is lengthened so that the body is entirely immersed in the water. As a result, equilibrium is destroyed. What should be done to restore equilibrium?

167. A vessel containing water is in equilibrium on a beam-balance. Will equilibrium be affected if you put your finger into the water without touching the bottom of the vessel (Fig. 86)?

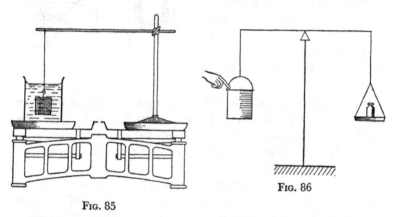

FIG. 85   FIG. 86

168. A body immersed in a liquid is balanced on scales (Fig. 87). Will the reading of the scales be altered if the liquid be heated together with the body?

169. One of the many fallacious designs for a *perpetuum mobile* consisted in the following: a shaft is set in a slit in the side of a vessel containing liquid, the shaft's axis lying in the same plane as the vessel's side (Fig. 88); the shaft completely fills the slit so that no liquid runs out. The shaft can rotate about its own axis. An upthrust acts on the half of the shaft which is immersed in water (according to Archimedes' principle) which ought, it would appear, to cause the shaft to rotate in a counter-clockwise

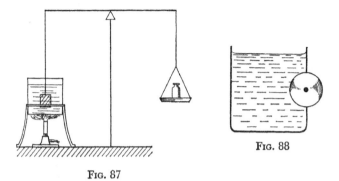

FIG. 88

FIG. 87

direction. This rotation, if it arose, ought to continue perpetually. Where is the fallacy here?

170. An open beaker is lowered into a vessel containing water, once bottom upwards, and once bottom downwards (Fig. 89), both times to the same depth. In which of the two cases will the work which must be expended on dipping the beaker into the water be greater? (The water does not spill out of the vessel, nor does it flow over into the beaker when dipped in bottom downwards.)

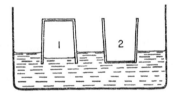

FIG. 89

171. Why is the waste-pipe of a washbasin connected to the vertical drainage-pipe by a siphon-tube (Fig. 90) and not go straight there?

172. The following experiment is suggested to prove the compressability of liquids. A round flask with capillary tube, containing the liquid under examination (Fig. 91), is soldered to an exhaust pump. After all the gases dissolved in the liquid (chiefly air) have been removed by simultaneously heating the flask and pumping out air, the flask is unsoldered from the pump at the top part of the capillary tube. When the flask has come to the same temperature as the surrounding medium, the end of the

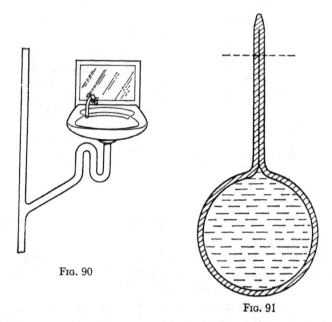

Fig. 90

Fig. 91

capillary tube is cut off and the reduction in the level of the liquid in the capillary tube is observed. Is this experiment valid?

173. Will a hydraulic press work if its cylinder be filled with gas instead of liquid?

174. When a steam boiler in which the pressure of steam is 10–15 atm blows up, considerable damage is caused; yet when a hydraulic press, in which the pressure is much higher, blows up, not very great damage results from the explosion. Why?

175. Why are factory chimneys built so tall? And which chimneys are better—steel or brick ones?

176. A pipe is introduced into the smoke-box of steam-locomotives, through which a stream of steam shoots up (Fig. 92). Why is this done?

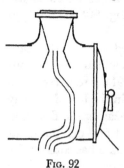

Fig. 92

177. A pipe is set in the bottom of a vessel containing water so that it can rotate freely about its own axis. Two nozzles are soldered to the upper end of the pipe and bent as shown in Fig. 93. The whole system is a sort of inverted Segner's wheel. What will happen to the pipe when water flows out of the vessel through the pipe?

178. Why does a sinking ship often overturn as it settles in the water?

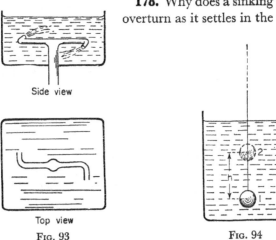

Side view

Top view

FIG. 93

FIG. 94

179. A metal ball, suspended from a thin thread, is lowered into a beaker containing a liquid (Fig. 94). The ball is then raised through a height h. Its potential energy will clearly now have increased by mgh, where m is the mass of the ball. On the other hand, a volume of liquid equal to the volume of the ball v, will move downwards from position 2 to position 1, i.e. its potential energy will decrease by $v\rho gh$, where ρ is the density of the liquid. The potential energy of the whole system (that is, ball-liquid) has been altered. How can Archimedes's law be deduced from these considerations about energy?

180. A vessel of the shape shown in Fig. 95 is filled with water and caused to rotate. What will happen if the cork which closes the opening at A (which lies on the axis of rotation) is taken out?

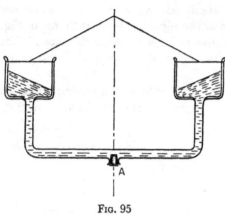

FIG. 95

181. Determine the surface-configuration of a liquid contained in a vessel which slips without friction down an inclined plane. (Fig 96).

182. A vessel containing a liquid is standing on a trolley. The trolley is moving in a horizontal line under acceleration. Determine the surface configuration of the liquid.

183. Water is poured into a U-tube. The tube is caused to rotate with an angular velocity of ω about an axis passing through one of the limbs of the tube (Fig. 97). How can the level of water in the two limbs be found?

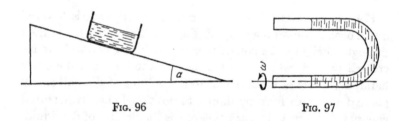

FIG. 96   FIG. 97

184. A vessel containing water is suspended from strings to form a pendulum. To its sides are attached two extended, soft springs (Fig. 98). What will happen to the level of the liquid in the vessel when the pendulum starts to swing?

185. What is the principle of the rubber bulb of an ordinary atomizer? What is the purpose of the second rubber bulb? Why is the first one made of thick rubber and the second of thin? Why is the second bulb enclosed in a net (Fig. 99)?

186. If a liquid is poured into a glass and small bodies heavier than this liquid be put into it, then these bodies will move towards the sides of the glass when the latter is rotated. How can we explain the well-known fact that tea-leaves, which move

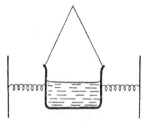

Fig. 98

Fig. 99

towards the sides of the glass when the tea is stirred, collect in the middle of the bottom of the glass when stirring stops?

187. On some railways steam-locomotives take on water without stopping. For this purpose a pipe bent at a right angle is used, being lowered into a trough between the rails while the train is moving (Fig. 100). What must be the speed of the train v, if the water is to be raised through a height of $h = 3$ m?

188. An opening S is made in a vessel containing liquid. The opening is small by comparison with the height of the column of liquid. In one case the opening is closed with a disk and the force of the liquid's pressure F_1 on the disk is measured, when the height of the column of liquid is h (Fig. 101a). In another case the same vessel stands on a trolley, with the opening unstopped, and the force of recoil F_2 is measured with the water flowing out at a moment

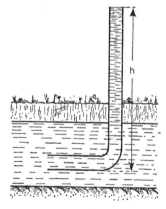

Fig. 100

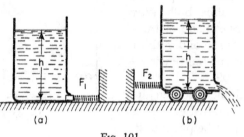

Fig. 101

when the height of the column of liquid is the same as in the first case (Fig. 101b). Will the forces F_1 and F_2 be equal?

189. If a finger is placed under a tap which is turned fully on, so that only a small opening is left, the water will gush out with a greater velocity than when the tap is free to flow normally. Why does this happen?

190. Is it better for a plane to take off into the wind or with the wind?

191. What conditions must obtain for a plane to be able to fly upside down?

192. Two planes of the same type fly, the first along arc ABC, the second along arc ADC (Fig. 102). Both arcs lie in a vertical plane and they are of the same length. Which plane will have the greater velocity at C, given that they both have the same speed at A and that their engines develop a constant power, which is the same for both planes?

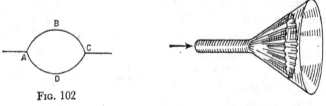

Fig. 102

Fig. 103

193. Is it possible to blow a paper filter fitted into a funnel out of the funnel, when blowing from the narrow end (Fig. 103)?

194. To separate thin pieces of paper from one another (as, for instance, in books of Underground tickets), it is enough to blow sideways on at the edges of these pieces. How is this phenomenon to be explained?

IX. Heat and Capillary Phenomena

FIG. 104

195. The experiment demonstrating the expansion of metal under heat is well known: a metal ball which passes through a metal ring gets stuck when it is heated (Fig. 104). What will happen if the ring, instead of the ball, is heated?

196. Why are calorimeters made of metal and not of glass?

197. Which thermos-flasks are best, given the same height and capacity, ones of square or of circular cross-section?

198. There are two layers of water in a calorimeter, the lower one colder, the upper one hotter. Will the overall volume of the water be altered if the temperatures are evened out?

199. Two teapots, one made of copper and weighing 200 g, the other of china and weighing 300 g, both have a capacity of 500 g of water. Tea brews better, the higher the temperature of the water. Which teapot would produce a better brew, if it were possible to neglect surface cooling of the teapots, when the water is poured from the kettle into teapots at room temperature (20°C approx.)? (The specific thermal capacity of copper is 0·095 and of china is 0·2.) How does it help to heat the teapot first with water from the kettle? Which teapot is better in fact, when the teapots are subject to surface cooling?

200. A piece of metal and a piece of wood are at the same temperature. Why does the metal feel colder to the touch than the wood, when cold, and hotter than the wood when hot? At what temperature will the metal and the wood both feel as though they are at the same temperature?

201. A plate composed of welded sheets of copper and iron is

connected to an electrical circuit
as shown in Fig. 105. Describe
what will happen if a fairly
strong current be passed through
the circuit.

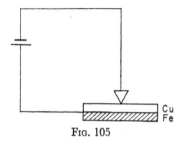

Fig. 105

202. It is well known that to
take a man's temperature with
a thermometer 5–10 min are
needed; but to shake the
mercury down when the thermometer is taken out sometimes
requires only a few seconds. Why is this?

203. What properties of red copper make it an exceptionally
suitable substance for soldering irons? Is there any other substance which has such valuable properties for the purpose?

204. Two cylinders of identical
dimensions but of different mat-
erials are welded together at their
butt-ends (Fig. 106). The thermal
capacity of cylinder A is twice as

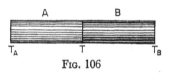

Fig. 106

great as that of cylinder B, but its thermal conductivity is only
half that of B. One of the free ends is heated and the other is
cooled, so that constant temperature is
maintained at each end. Will the overall
quantity of heat flowing through the cylinders depend on whether it is A's end that
is heated and B's cooled, or vice versa? (Do
not take into account the loss of heat through
the curved sides of the cylinders.)

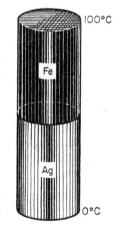

Fig. 107

205. Two cylinders of the same dimensions, one of iron, the other of silver, stand
one upon the other (Fig. 107). The lower
end of the silver one is kept at a temperature
of 0°C, and the upper end of the iron one is
kept at a temperature of 100°C. The thermal
conductivity of silver is eleven times greater

than that of iron. What is the temperature of the ends which are in contact with each other, if we assume that no heat escapes into the surrounding medium through the side surfaces of the cylinders?

206. The calorific value of pinewood is higher than that of birch. Why is it said that birch wood 'burns hotter'?

207. Is the power of the Dneper hydroelectric station sufficient to raise the temperature of the water that passes through its turbines to boiling-point? The temperature of the water in the river is 20°C.

208. A thin sheet of mica is placed on a horizontal copper plate, and a heated metal cone is laid on top of that. The cone begins to roll round its own apex on the mica. Why? What would happen if there were not a copper plate, but a sheet of glass under the mica?

209. One often sees a housewife who turns up the flame under a saucepan in order to make the contents boil faster. Is this the right thing to do?

210. What is the temperature of the water in a reservoir under ice?

211. The temperature of 0°C is both the temperature at which ice melts and that at which water freezes, as is well known. What will happen if we put a piece of ice at temperature 0°C into a vessel containing water also at 0°C?

212. Equal quantities by weight of water at + 50°C and of ice at −40°C are mixed together. What will be the final temperature of the mixture?

213. To heat a saucepan of water as quickly as possible, it is always put above the source of heat (e.g. it is put on an electric ring). Wishing to cool a saucepan of water to room temperature as quickly as possible a housewife puts it on ice. Is this the right way?

214. Will water boil if it is in a saucepan floating in another saucepan containing boiling water?

215. A glass of hot water is to be cooled as much as possible in

10 min. Which is better: to put a piece of ice in and then leave the water to cool for 10 min, or to leave the water to cool for 10 min and then put the same amount of ice in?

216. The temperature of water in open reservoirs (or ponds, lakes or rivers) is almost always lower in summer weather than that of the surrounding air. Why?

217. Can water be made to boil without heating it?

218. A certain quantity of water is put into two hollow glass spheres which are connected by a tube, and the air is then pumped out and the whole system is sealed. If all the water is poured into the upper sphere and the lower one is put in a vessel containing liquid air (Fig. 108), after a certain time all the water in the upper sphere will freeze, although the sphere remains at room temperature throughout. Explain this phenomenon.

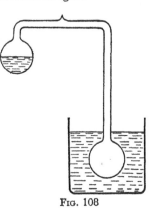

Fig. 108

219. Mercury boils at a temperature of +367°C. How then can mercury thermometers be used to measure temperatures of up to 550°C?

220. Two identical vessels filled with hydrogen are connected by a tube in which there is a small column of mercury (a liquid cork, Fig. 109). In one vessel the hydrogen is at 0°C, and in the other at +20°C. Will the column of mercury be displaced if both vessels are heated through 10 degrees?

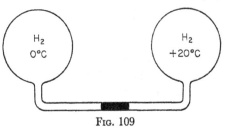

Fig. 109

221. A jar containing cloudy water is left overnight on the window-ledge. By morning cloudiness remains only on the side of the jar turned towards the room. At what time of year is the experiment performed?

222. In a room at temperature $+20°C$ a humidity reading of 40 per cent is taken. At the same time the humidity outside, where the temperature is $0°C$, is found to be 80 per cent. If an outside window in the room be opened, which way will the water-vapour move: from the room outside, or from outside into the room?

223. Two soap-bubbles of different diameters are blown at the two ends of a bent glass tube. Will the diameters of the bubbles alter if the tap K is closed (Fig. 110)?

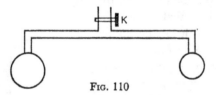

FIG. 110

224. A glass vessel has a small hole in the bottom of radius r. To what height can a liquid which does not wet glass be poured into the vessel without running out?

225. A capillary tube of radius R is dipped into a liquid so that only a length L projects above the surface of the liquid, L being less than the height to which the given liquid will rise up a capillary tube of the given cross-section with full wetting. In which direction will the meniscus curve, and what will be its radius, if the coefficient of surface tension of the liquid is σ and its density is ρ?

226. In a MacLeod manometer (Fig. 111), which is used for measuring very small pressures, there are two tubes parallel to the sealed capillary tube, one being a similar capillary tube and the other a wide tube. Why are both these tubes necessary?

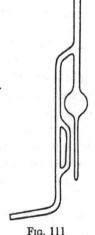

FIG. 111

X. Electricity

227. Why do birds fly off a high-tension wire when the current is switched on?

228. How can a charged conductor give up all its charge to another, insulated, conductor?

229. We know that a charged ball attracts a piece of paper. How will the force of attraction be altered if a metal sphere be placed round (a) the charged ball (b) the piece of paper?

230. A small metal ball is charged to a potential of $+1$ V. It is introduced into a large hollow metal sphere charged to a potential of $+10,000$ V and comes into contact with the inside surface of the sphere. The ball's charge passes to the sphere. Explain the apparent contradiction in the passing of a positive charge from a body at lower potential to another body at higher potential, when exactly the opposite ought to take place.

231. N identical drops of mercury are charged simultaneously to the same potential V. What will be the potential V' of the large drop formed by combining these drops? (Assume that the drops are of spherical shape.)

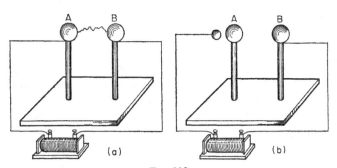

Fig. 112

232. A spark discharge occurs between two spheres A and B, which are connected to an electrostatic machine and mounted on an insulating stand (Fig. 112a). In time a 'leak' occurs in the stand, i.e. it begins to conduct electricity to an insignificant degree, and therefore the spark discharge stops. Why is it that when a subsidiary spark gap is set up between the electrostatic machine and sphere A (as shown in Fig. 112b), the spark discharge can be re-established between A and B?

233. We know that the interaction of two charges is less in water than in air. It would appear that this could be utilized to create a *perpetuum mobile* in the following way: take two opposite charges at a and b (Fig. 113) and bring them together in the air, then drop them simultaneously into water, move them apart under water, then raise them simultaneously into the air to their

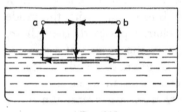

FIG. 113

previous positions and repeat the process from the beginning. Then the work stored up when the charges come together is greater than that expended on moving them apart, since the force of electrical interaction in air is greater than in water. Where is the fallacy?

234. A capacitor is connected up with an accumulator. If we move the plates of the capacitor apart, we overcome the force of electrostatic attraction between the plates and consequently we do positive work. On what does this work go? What happens to the energy of the capacitor?

235. A flat capacitor, in which the plates are large by comparison with the distance between them, is connected to a source of constant voltage. Will the strength of the electric field inside the capacitor be altered if the space between the plates be filled with a dielectric?

236. The plates of a flat capacitor are connected to a galvanometer (Fig. 114). One of the plates is earthed. A positive charge is passed between the plates. What will be the reading of the galvanometer?

237. Two metal spheres of the same radius are placed far apart, i.e. the distance between their centres is much greater than their radius. What is the capacity of the system formed by these two spheres?

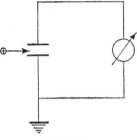

FIG. 114

238. Is it possible to measure the voltage of an a.c. circuit with an electroscope?

239. At several points on tramlines there are automatic signals saying 'Beware of Trams'. The signal lights up before the tram comes up to it and goes out when the tram passes. What system would allow this signal to be switched on?

240. It is required to light a corridor with a lamp hung in the middle of the corridor and so arranged that it can be switched on from either end of the corridor. What switch-system will allow this to be done?

241. When wiring for lighting is installed in a house, using alternating current of 220 V from a three-phase circuit in which the voltage between two phases is 380 V, two wires are used: 'neutral' and 'phase'; it is not allowed to fit fuse plugs to both wires, and so fuse plugs are put on only one, the 'phase' wire. Why?

242. Two electric bulbs, made for 120 V and 40 W, one with a metal filament, the other with a carbon one, are connected in series to a circuit of 120 V. Which of the filaments will incandesce more?

243. How is the incandescence of two electric bulbs

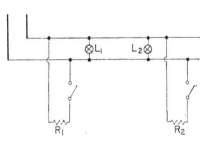

FIG. 115

L_1 and L_2 in a flat affected by electrical heating appliances being switched on and off (R_1 and R_2) if the appliances and the lamps are connected as shown in Fig. 115?

244. Three lamps are connected as shown in Fig. 116. All the lamps are of the same power and intended for a current of 120 V. How will the current in lamps L_1 and L_2 be affected if switch K is closed.

245. An electric kettle begins to boil 15 min after being switched on. The heating element consists of a coil of wire 6 m long. How should the heating element be adapted so that the kettle begins to boil 10 min after being switched on? Neglect loss of heat to the surrounding atmosphere.

246. Find the electrical resistance of a homogeneous wire frame in the shape of a regular hexagon with two diagonals linked together at O (Fig. 117). The current is led into the frame at the mid-points A and B of opposite sides of the hexagon. The resistance of one side of the hexagon is R.

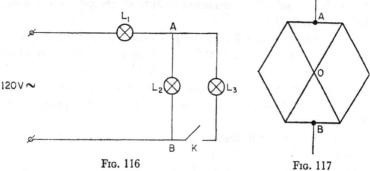

FIG. 116   FIG. 117

247. Two cells with the same e.m.f. E and different internal resistances r_1 and r_2 are connected in series to an external resistance R (Fig. 118). Can a value for R be selected such that the potential difference at the terminals of the first cell should be zero?

248. Two lamps with resistances of r and R at full incandescence, R being greater than r, are connected in series in a

lighting circuit. Which of the lamps will shine more brightly?
Both have wolfram filaments.

249. Is it possible for two cells of respective e.m.f. E_1 and E_2 and
respective internal resistance r_1 and r_2 to produce a weaker current
when connected in series to an external resistance R than one of
the cells by itself, connected to the same resistance?

250. Three identical batteries are connected as shown in Fig.
119. What will the voltmeter (V) register if it is connected in
parallel with one of the cells at A and B? Neglect the resistance of
the connecting wires. Will this result be altered if the number of
batteries be changed and the voltmeter's points of connection
left the same?

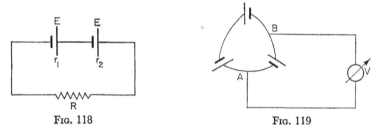

Fig. 118   Fig. 119

251. Is it possible to determine the e.m.f. of a battery by connecting up instruments as shown in Fig. 120?

252. How can the resistance of a galvanometer which is normally connected on the diagonal of a Wheatstone bridge be
measured on the bridge without using another galvanometer
(Fig. 121)?

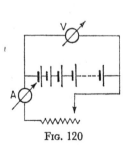

Fig. 120

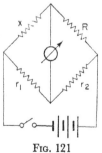

Fig. 121

253. In what conditions will the strength of current in a wire be the same for connection in series and connection in parallel of n identical cells?

254. Two batteries with e.m.f. of E_1 and E_2 respectively are connected in a circuit as shown in Fig. 122. Resistances are chosen so that the ammeter (A) shows an absence of current. What will be the reading of the voltmeter (V)?

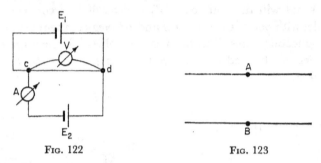

FIG. 122   FIG. 123

255. Two arbitrary points A and B are chosen on a direct current two-wire system, one point on each wire (Fig. 123). How can we find in which direction is the source of current with the help of a voltmeter and a magnetic needle?

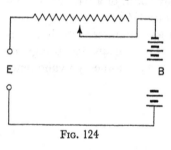

FIG. 124

256. A battery B, consisting of 60 accumulator cells is charged from a d.c. source E of 115 V (Fig. 124). The charging current must be of 2·5 A, the e.m.f. of each cell at the beginning of charging is 1·2 V and the internal resistance of each cell 0·02Ω. What must be the resistance of the rheostat connected between the source and the battery?

257. Two identical shunt-wound motors are each connected to a circuit of voltage V. One motor rotates freely, the other does work. Which of them will heat up more quickly?

258. In supplying an electric furnace P with direct current, the required temperature is maintained when the hot-wire ammeter (*A*) registers 5 A (Fig. 125a). Will the same temperature be maintained when the furnace is supplied with alternating current and the same ammeter reads 5 A? What will happen if the hot-wire ammeter is replaced by an electromagnetic one?

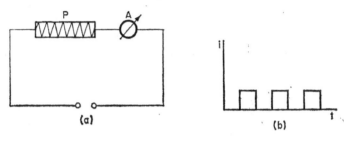

FIG. 125

259. Can temperature changes affect the readings of ammeters and voltmeters?

260. Two wires *MACN* and *MBDN*, of the same length but of different resistances R_1 and R_2 are connected as shown in the diagram of Fig. 126. How should contacts *A*, *B*, *C* and *D* be arranged so that there should be no current passing through wires *AB* and *CD*? Will current pass through *AB* and *CD* with this arrangement of contacts if two points *E* and *F* on these wires are connected?

261. A constant potential difference is applied to the ends of a

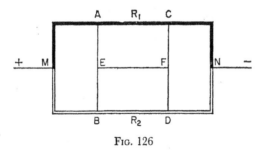

FIG. 126

graphite rod, whose resistance decreases with a rise of temperature. When will the amount of heat given off by the rod be
greater: when it is bare, or when it is covered with asbestos?

262. A circuit consisting of two capacitors of the same capacitance connected in series is connected to a plug supplied with
current from a three-phase lighting system.

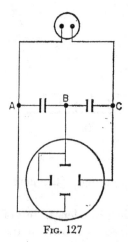

The current is transmitted from the plates
of one capacitor to one pair of plates of an
electron oscillograph and the current from
the plates of the other capacitor to the
oscillograph's other pair of plates. A line
appears on the oscillograph screen, inclined at an angle of 45° to the horizontal.
If point A is now connected to earth (Fig.
127), the picture on the screen remains the
same. If point B is earthed, the line on the
screen becomes horizontal. If point C is
earthed the fuse in the plug burns out.
What is the explanation of all this?

Fig. 127

263. A chain of Christmas-tree lights is
composed of 40 pocket-torch bulbs connected in circuit and supplied from the town grid. After one bulb burns itself out the
remaining 39 bulbs are again
connected in series and plug-
ged in to the town network. In
which case will the room be
more brightly lit, when there
are 40 bulbs or only 39?

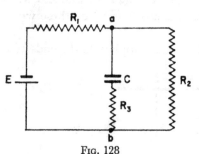

Fig. 128

264. A battery of e.m.f. E
is connected to a system as
shown in Fig. 128. To what
voltage will the capacitor C
be finally charged? (Neglect the battery's internal resistance.)

265. An accumulator battery is connected to a container full of
acid solution. On discharging, an amount of inflammable gas is

obtained, such that upon its combustion 35 per cent of the energy expended on charging the accumulator battery is given off. If several containers of acid solution instead of one are connected in series, then of course more time will be required for the same amount of electricity to flow through them than in the case of the single container. But the amount of liberated material depends, according to Faraday's law, only on the amount of electricity that passes through the electrolyte. Therefore after a sufficiently long period the same amount of inflammable gas will be liberated in each container as when only one container of acid solution was connected. Then, upon combustion of all the inflammable gas that has formed in all the containers, energy will be liberated which will far exceed the energy expended on charging the accumulator battery. In other words, the law of conservation of energy seems to have been broken. Explain the difficulty.

266. During electrolysis the depositing of a substance at the cathode depends on the positive ions. The total current in the electrolyte is composed of two currents: that of the positive ions (I_+) and that of the negative ions (I_-), which move in opposite directions. Then why is it that the amount of a substance deposited at the cathode is calculated from the total current, i.e. from the sum of the currents I_+ and I_-, and not from current I_+ only?

267. Tramlines are supplied by direct current. The overhead wire is connected to the positive pole of the dynamo and the rails to the negative pole. Why not the other way about?

268. An electric current in a metal conductor is the movement of free electrons, which collide with the ions of which the crystalline structure of the metal is composed; the electrons thereby give up to the ions all the momentum which they had acquired before collision.

Then why does a metal conductor through which current is passing not experience the action of mechanical forces in the direction of motion of the electrons?

269. In which case will electrons reach the anode with the

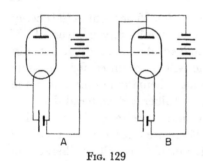

FIG. 129

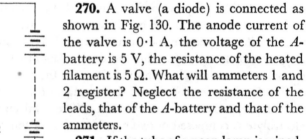

FIG. 130

greater velocity: when a valve is connected as shown in diagram A or as in diagram B (Fig. 129)?

Consider two cases: (1) when the internal resistance of the anode battery can be neglected, (2) when the anode battery has a high internal resistance.

270. A valve (a diode) is connected as shown in Fig. 130. The anode current of the valve is $0 \cdot 1$ A, the voltage of the A-battery is 5 V, the resistance of the heated filament is 5 Ω. What will ammeters 1 and 2 register? Neglect the resistance of the leads, that of the A-battery and that of the ammeters.

271. If the tube of a neon lamp is wiped, it can be seen to gleam for a short period. What is the explanation of this phenomenon?

272. Two magnetic needles mounted on vertical pivots are so arranged that opposite poles are at a short distance from one another (by comparison with the length of the needles). If the needles are drawn to one side through a small angle and then released (Fig. 131a), they will begin to oscillate. Will the period of oscillation be altered if the needles are drawn aside not to

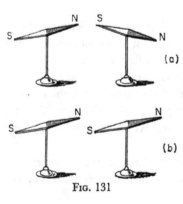

FIG. 131

the same side, but to opposite sides (Fig. 131b)?

273. How can we find whether a fret-saw shaving is magnetized without using any other body?

274. You are given a homogeneous wire frame in the shape of a cube (Fig. 132). A constant voltage is led into the opposite corners of one diagonal of the cube. Currents pass along the edges of the cube. What will be the strength of the magnetic field at the centre of the cube?

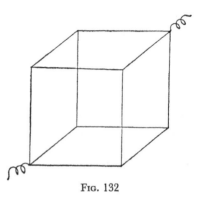

Fig. 132

275. A soft, thin, metal filament runs along a rigid wire, through which is passed alternating current from the city grid. In one case a.c. is also passed along the filament from the grid, and in another case d.c. is passed along it. What will happen to the filament in both cases?

276. A solenoid in the shape of a toroid is placed in the plane of the magnetic meridian. A compass is placed at the centre of the solenoid (Fig. 133). How will the compass needle behave if d.c. is passed through the solenoid?

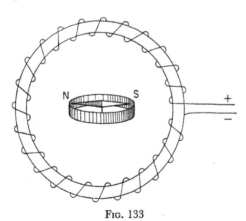

Fig. 133

277. A magnet is brought close to a rotator made of metal wires. Alongside the magnet and below the rotator is placed a burner which heats one of the rotator's wires (Fig. 134). What will happen?

278. Many physics textbooks, when dealing with the working of the electromagnetic telegraph, give, more or less, the system shown in Fig. 135. The arrangement of instruments at the transmitting and receiving stations is exactly the same. Clearly the transmission of telegrams according to this arrangement takes place without leaving any copy at the transmitting station

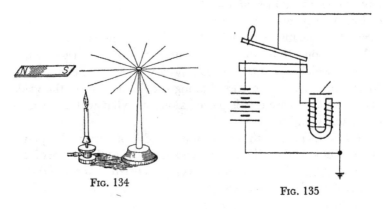

FIG. 134

FIG. 135

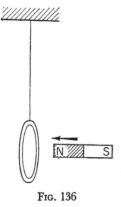

FIG. 136

and also simultaneous transmission of telegrams from both ends is impossible. With the aid of an additional electromagnet at each station, compose an arrangement in which both these drawbacks are obviated, i.e. allowing a copy of the telegram transmitted to be made at the sending station and also allowing simultaneous transmission of telegrams in both directions.

279. A copper ring is suspended in a vertical plane by a thread. A steel bar is passed through the ring in a horizontal direction, and then a magnet is similarly passed through (Fig. 136). Will the motion of the bar and the magnet affect the position of the ring?

280. In Thomson's experiment a coil consisting of a large number of turns of copper wire is wound round a steel rod. A massive ring of some metal of high conductivity (copper) is passed freely over the core. When the coil is connected to an a.c. circuit the ring jumps up (Fig. 137). Will the ring jump up if the coil is connected to a d.c. circuit?

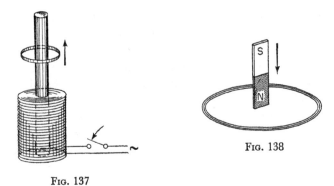

FIG. 137   FIG. 138

281. A straight permanent magnet falls through a metal ring (Fig. 138). Will the magnet fall with the acceleration of a freely falling body?

282. A coil *A* is connected to a voltmeter and another coil *B* is connected to an a.c. source (Fig. 139). How will the potential induced in coil *A* by the current in coil *B* be altered if a large copper sheet is placed between the two coils?

283. Two coils through which current is passing act on each other with a given force. How will this force be altered if both coils are passed freely over a closed steel core, all the lines of force of the magnetic field passing inside the core?

284. The famous English physicist Michael Faraday discovered the phenomenon of magnetic induction in 1831. Faraday looked long and patiently for this phenomenon, guided by the

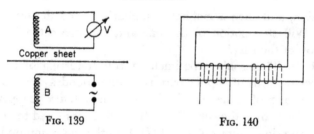

FIG. 139   FIG. 140

general idea of a connection between the phenomena of electricity and magnetism.

At the same time, and independently of Faraday, the Swiss physicist Colladon was working in the same direction, guided by the same idea. Colladon's experiment consisted in the following: the ends of a solenoid are connected to a galvanometer, which was taken into the next room in order to obviate the direct influence of the magnet.* Colladon moved a magnet into the solenoid and then went into the next room to see what the galvanometer read. What was Colladon's mistake? Why did he

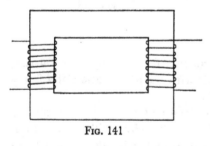

FIG. 141

not succeed in discovering the phenomenon of magnetic induction?

285. Two windings are passed over a closed steel core (Fig. 141). How can the number of turns in each winding be found, having at one's disposal an a.c. source, wires, and voltmeters of any sensitivity?

286. A closed steel core has on it two identical windings with an ohmic resistance which is far lower than its inductive resistance. One winding is connected through an ammeter (Fig. 142) to an a.c. source. Will the reading of the ammeter be altered if the ends

---

* In Colladon's time they used galvanometers in which a light magnetic needle was suspended inside a coil. The presence of currents in the coil was determined by the needle. In a galvanometer of this sort the effect of a magnet placed near the galvanometer cannot be avoided.

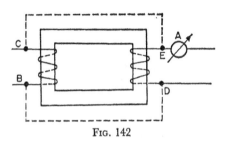

FIG. 142

B and *C* of the second winding are connected with points *D* and *E* on the first coil in such a way that the magnetic currents of both windings act in the same direction?

287. A straight wire begins to move with increasing velocity, crossing the lines of force of a homogeneous magnetic field which is directed at right angles to the plane of the paper (Fig. 143). In one case the ends of the wire are connected to an ohmic resistance *R*. In the second case a self-induction coil *L* is connected in series to the same resistance *R*. What does the work expended on moving the wire become in both cases? In which of the two cases will more work be done, given the same displacement of the wire?

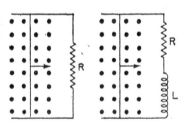

FIG. 143

XI. Optics.

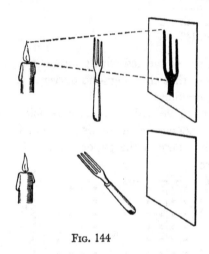

288. A fork is illuminated by a candle and casts a shadow on the wall. When the fork is in a vertical position, the shadow reproduces the shape

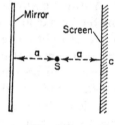

FIG. 144   FIG. 145

of the fork's prongs distinctly, but when it is in a horizontal position the shadow is blurred and the prongs are not visible (Fig. 144). Why?

289. A screen is placed at distance a from a point source of light S. How will the illumination at the centre of the screen be altered if a plane mirror be placed on the other side of the source at the same distance a from it (Fig. 145)?

290. What should be the least size of a plane mirror for a man to be able to see himself at full length when standing in front of it?

291. A photographer with an extending camera takes a photograph of a man, and then photographs the clouds floating across the sky. Should he increase or decrease the camera's extension?

292. A point of light lies between two plane mirrors which are parallel to each other. How many images of the point will be obtained in the mirrors?

293. A point of light A lies between two plane mirrors which are arranged at right angles to each other. How many images of the point can be seen in the mirrors?

294. Arrange two mirrors so that, whatever the angle of incidence, the incident ray and the resulting ray reflected from the two mirrors should be parallel to one another.

295. Arrange three mirrors so that they all intersect and so that, whatever the angle of incidence, the incident ray and the resulting ray reflected from the three mirrors should be parallel to one another.

296. In what medium can rays of light be curved?

297. An eclipse of the sun is being observed. At one moment the edge of the sun, the edge of the moon and the observer's eye are in the same straight line. How soon after this will the observer first see the sun's rays?

298. When the sun rises it often appears to be flattened. Why?

299. A man A stands to one side of a mirror (Fig. 146); a second man B approaches the mirror along the line perpendicular to it which passes through its centre. At what distance from the mirror will B be at the moment when A and B see each other in the mirror?

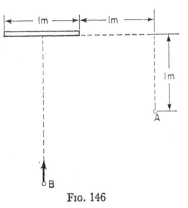

FIG. 146

300. A small square mirror S lies on a table. Of what shape will the sun's reflection from this mirror be on a screen E which is positioned in a vertical plane and at a reasonable distance from the mirror (Fig. 147)?

301. Of what shape should the front surface be of the cornea of

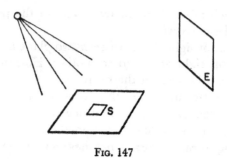

Fɪɢ. 147

an imaginary animal which could see distant objects equally well in the air and under water without any supplementary accommodation?

302. An image of a candle is obtained on a screen with the aid of a double convex lens. How will this image be altered if the lens be half covered by a piece of cardboard (Fig. 148)?

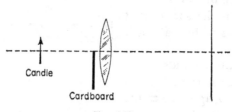

Candle

Cardboard

Fɪɢ. 148

303. A double convex lens of focal length f lies between a source of light and a screen. The distance between the source of light and the screen is less than $4f$. It is known that in these conditions it is not possible to obtain an image of the source on the screen, whatever the position of the lens. How can an image of the source be obtained on the screen with quite simple means and without moving either lens or screen?

304. Is it possible in certain circumstances to obtain a real image in the air with the aid of a double concave glass lens?

305. A ray of light falls on to a transparent homogeneous

sphere and passes into it.
After passing through the inside of the sphere it reaches
the sphere-air surface at A
(Fig. 149). Can total internal
reflection take place at this
point?

306. In Fig. 150 is de-
picted the path of a ray of
light BC after refraction in

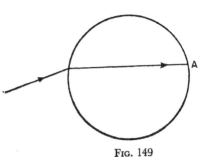

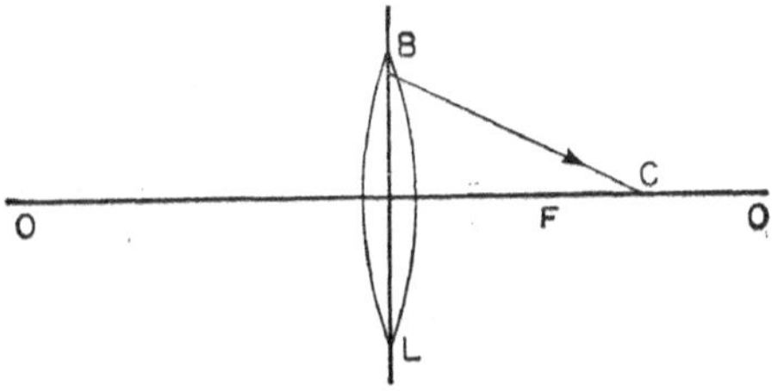

FIG. 149

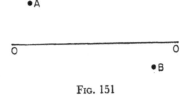

FIG. 150

a double convex lens L of principal focus F and of principal axis
OO. Find by construction the path of this ray before reaching
the lens.

307. With the help of a lens of principal axis OO an image B
is obtained of point A (Fig. 151). Where is the lens placed, of
what kind is it and where are its foci?

• A

O ———————————————— O

• B

FIG. 151

308. How should two converging lenses be placed so that a
parallel beam should become parallel once more after passing
through both lenses?

309. Where should a point source of light lie along the principal axis of a converging lens so that it is impossible to see the source and its image simultaneously from any point?

310. A disk whose plane surfaces are parallel is cut as shown in Fig. 152a; then the lenses so obtained are moved apart. What will happen to a beam of parallel rays falling on to the resulting system:

(a) from the side of the converging lens (Fig. 152b), (b) from the side of the diverging lens (Fig. 152c)?

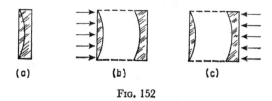

(a)   (b)   (c)

Fig. 152

Consider the cases when the distance between the lenses is less than the focal length and when it is greater than the focal length.

311. A broad beam of parallel rays of light, consisting of rays of two colours of the spectrum is dispersed in a liquid. How can the rays of the two colours be separated from each other with the aid of a thin transparent disk whose plane surfaces are parallel, if the coefficient of refraction of the disk's material is less than that of the liquid and if the value for this is different for the rays of the two colours?

312. In H. G. Wells's novel *The Invisible Man*, the hero of the book discovered a compound which, when he drank it, made him completely transparent to rays of light—and therefore invisible. In the novel the invisible man himself can see his surroundings, while remaining unseen. Can such an invisible man see?

313. Why do the windows of a house seem dark in daylight, i.e. darker than the walls of the house, even though the walls are painted some dark colour?

314. If you look at a luminous advertisement, for example at a

neon sign, the red letters always seem to stick out in front of the
blue or green ones. How is this to be explained?

315. Why is dry sand light, while wet sand appears dark?

316. To protect oneself from the heat of a red-hot stove, it
helps to place a sheet of glass in front of it, and not a sheet of
ebonite, since glass is not highly transparent to heat (infra-red)
rays, while ebonite is transparent to them. Then why is it that
glass and not ebonite is used for hot-houses?

317. The sun's rays are made to converge by a concave mirror
and are directed into a hollow sphere after passing through a
small hole in its surface (Fig. 153). The walls of the sphere do not

Fig. 153

conduct heat. Is it possible, by increasing the dimensions of the
mirror indefinitely, to raise the temperature inside the sphere
without limit?

Solutions

I. Kinematics

1. Since the first passenger found that the train's speed was 31·2 km/hr, this means that in the 3 min ($\frac{1}{20}$ hr) the train travelled $31 \cdot 2 \times \frac{1}{20} = 1 \cdot 56$ km. From this it is clear that he obtained the distance 1·56 km, or 1560 m by multiplying 10 by 156, whereas he ought to have multiplied by 155, since the first rail is passed on the second click. The second passenger found the train's speed to be 32 km/hr. Therefore, according to his calculation, the train travelled $32 \times \frac{1}{20} = 1 \cdot 6$ km or 1600 m. From this it is clear that he multiplied 50 by 32, whereas he ought to have multiplied by 31 since the first section of 50 m of track is passed on the appearance of the second telegraph-pole. So both passengers made a mistake in beginning their counting from nought instead of from one. The real speed of the train was $155 \times 10 \times \frac{1}{20} = 31,000$ m/hr $= 31$ km/hr, or alternatively $50 \times 31 \times \frac{1}{20} = 31,000$ m/hr $= 31$ km/hr.

2. A steamer setting out from A will meet first those steamers which have already set out from B and are *en route*, and, second, those which will leave B during the steamer's trip from A. At the moment of the latter's departure there are 12 steamers already *en route*, counting the one which has just left B (but not counting the one which has just arrived at A). Besides this, in the 12 days during which the steamer is on its way from A, 11 steamers will leave B (not counting the one which is leaving B as A arrives). Thus each steamer meets $12 + 11 = 23$ steamers in the open sea.

The solution can be explained by a diagram of the steamer's movements (Fig. 154). Letting the axes of the graph represent distance and time, plot the movement of every steamer leaving B and of one steamer leaving A.

From the intersections of these lines it is at once plain that each steamer meets 23 steamers in the open sea and that two additional meetings occur—in port A, at the moment of departure, and in port B, at the moment of arrival.

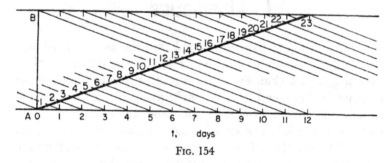

Fig. 154

3. Let the unknown time be t sec. Then the car will have time to move a distance of vt in this time, where v is the velocity of the car. During this period the image must not move more than 0·1 mm. The ratio between these two values must clearly be the same as the ratio between the sizes of the object and its image, i.e. $300 : 1·5 = 200$. Therefore

$$t \times \frac{36 \times 100,000}{3600} \div 0·01 = 200,$$

$$t = \frac{200 \times 0·01}{1000} = 0·002 \text{ sec.}$$

4. The usual answer to this question is that the average speed is 35 km/hr. But this is not true. It would be true if the car travelled at the speeds given for equal periods of time. But it is clear from the conditions of the question that the car takes two different periods of time (t_1 and t_2) to travel the same *distance* s in the two directions, since:

$$s = v_1 t_1 = v_2 t_2.$$

Therefore $s = 40 t_1 = 30 t_2$, hence

$$t_1 = \tfrac{3}{4} t_2.$$

The average speed

$$v_{\mathrm{av}} = \frac{2s}{t_1 + t_2} = \frac{60t_2}{\frac{3}{4}t_2 + t_2} = 34 \cdot 3 \text{ km/hr.}$$

5. Suppose that the boy throws n times a second. Then the time of flight of each ball upwards $t = 1/n$ sec. The time of rise equals the time of fall. But the distance and time of fall are connected by the formula

$$s = \frac{gt^2}{2} = \frac{g}{2n^2}$$

Therefore the height equals

$$s = \frac{g}{2 \times 2^2} \approx \frac{9 \cdot 8}{8} \approx 1 \cdot 23 \text{ m.}$$

6. Both stones move relative to the earth with the same constant and uniform acceleration g. Clearly one stone moves uniformly in relation to the other, and the constant speed of the first stone in relation to the other is equal to that speed which the first stone acquires in 1 sec, i.e. in the period that elapses between the two moments at which the stones start falling.

It is not difficult to carry out the necessary calculation.

The distance travelled by the first stone is found from the equation

$$s_1 = \frac{gt^2}{2}.$$

The distance travelled by the second stone from the equation

$$S_2 = \frac{g(t-1)^2}{2}.$$

The distance between the two stones increases with the lapse of time according to the formula

$$s_1 - s_2 = gt - \frac{g}{2},$$

i.e. the first stone moves uniformly in relation to the second stone with a velocity numerically equal to g.

7. The aeroplanes are moving relative to each other at a speed equal to the sum of their speeds, i.e. at a speed, v, of 400 m/sec. Between the firing of any two rounds a period of time, t, elapses $= \frac{1}{900}$ min $= \frac{1}{15}$ sec. The distance between the bullet-holes must equal the relative distance travelled by the second aeroplane during this time, i.e.

$$s = vt = \frac{400}{15} = 27 \text{ m approx.}$$

Since the length of the fuselage of an aeroplane rarely exceeds 27 m, not more than one bullet can normally hit the aeroplane in the given conditions of firing.

As a result of air-resistance every bullet will require a greater

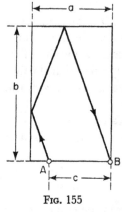

FIG. 155

length of time to traverse the distance between the two aeroplanes. But every bullet will be delayed by the same amount. Therefore the interval of time between the arrival at the target of any two consecutive bullets remains $\frac{1}{15}$ sec as before, and the distance between the bullet-holes must, as before, equal 27 m.

8. Let us resolve the velocity v imparted to the ball into components parallel with the sides of the table and consider the path of a ball as shown, for example, in the diagram (Fig. 155). We obtain two equations, evident from the diagram:

$$\frac{2a - c}{t} = v \cos \alpha, \qquad \frac{2b}{t} = v \sin \alpha.$$

From these equations we get:

$$\cot \alpha = \frac{2a - c}{2b},$$

i.e. we find angle α, at which the ball must be struck. The value for the velocity v which is imparted to the ball plays no part at all.

9. On impact all the balls have the same component velocity parallel to the short side of the table. In rebounding from the short sides the value of this component velocity does not alter, and therefore, since the tables are of equal width, all three balls will reach the opposite side of the table at the same moment (regardless of whether they have first struck the short sides or not). In rebounding from the long side opposite, the component velocity under consideration changes sign, but remains constant in size; all the balls therefore return to the

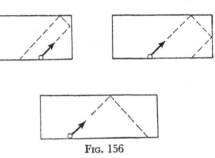

Fig. 156

side from which they started at the same moment (Fig. 156).

10. The speed of filling (i.e. the amount of liquid which falls into the bucket during unit time) will not change, for although

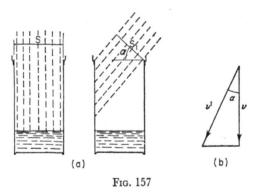

(a)   (b)

Fig. 157

the area of the cross-section of rain falling into the bucket decreases ($S_1 = S \cos \alpha$, Fig. 157a), the velocity of the drops not only changes direction but also increases in magnitude ($v' = v/\cos \alpha$, Fig. 157b). In other words, the speed at which the

D

bucket fills up depends only on the vertical velocity of the drops, which is not altered by the wind.

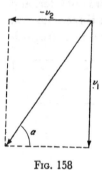

FIG. 158

11. Instead of the trolley's motion to the right relative to the drop, we can consider the drop's motion to the left with the same velocity relative to the trolley. The resultant velocity of the drop will then be compounded of two velocities at right angles to each other, v_1 and v_2 (Fig. 158). The problem requires that the resultant velocity should be parallel to the tube's axis, i.e. that it should be at an angle of α to the horizontal. This angle may be found from the equation

$$\tan \alpha = \frac{v_1}{v_2} = \frac{60}{20} = 3.$$

Hence $\alpha = 71° 35'$ approx.

12. Relative to the bucket (which floats downstream) the boat's speed upstream and downstream must be the same. Thus the boatman also spends 40 min on returning to meet the bucket. Further, the speed of the boat downstream relative to the bank is 25 m/min (since it takes 40 min to travel 1 km to point A). After meeting the bucket, the boat spends 24 min on returning downstream, so the meeting place is 600 m above A, or 400 m from the place where the bucket is lowered into the water. Thus the bucket travels 400 m in the 80 min between being lowered and picked up, i.e. the speed of the current is 5 m/min and that of the boat relative to the water is 20 m/min.

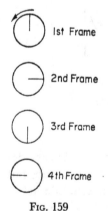

FIG. 159

13. This phenomenon is to be observed if the speed of change of frames in the film is very slightly greater than the speed of revolution of the car-wheels: then, in the time taken for one frame to succeed another, the wheels have time to perform rather less than one revolution. For example, in the case illustrated in Fig. 159, the wheel performs three-quarters of a revolution in a counter-clockwise direction. But to us it appears to have performed a quarter-turn in a clockwise direction during this period. If the wheel has time to perform one full revolution during each change of frame, the wheel will seem to stay still.

14. If the disk of a stroboscope with two holes performs N revolutions per second, then the total number of flashes per second will be $2N$ and the period between two flashes will be

$$t = \frac{1}{2N}.$$

The velocity of the drops at the point illuminated by the stroboscope will be

$$v = \sqrt{2gh} = \sqrt{2 \times 980 \times 22 \cdot 5} = 210 \text{ cm/sec.}$$

Since the velocity of the drops is great and the distance between any two of them is small ($s = 2$ cm), we can assume that each drop traverses this distance with constant velocity. Then the time taken by a drop to move through distance s is

$$t_1 = \frac{s}{v} = \frac{2}{210} \text{ sec.}$$

If the drops are to appear to be stationary, then $t_1 = t$, i.e.

$$\frac{1}{2N} = \frac{2}{210}.$$

Thus

$$N = \frac{210}{4} = 52 \cdot 5 \text{ rev/sec.}$$

If the drops have time to traverse a distance which is two, three, etc., times greater than the distance between the drops, then they will also be illuminated at the same point and will appear motionless. Therefore the number of revolutions can be any whole number of times less than 52·5 rev/sec. But the fewer the revolutions, the longer the drops will be illuminated (given a fixed size of hole), and the more noticeable will be each drop's displacement during this time. Therefore the picture will be less clear.

15. We shall see a circle of light where the illuminated apertures succeed one another at intervals of $\frac{1}{16}$ sec or less. Since the apertures are 1 cm apart, it is clear that their speed must be $v = s/t = 1$ cm/$\frac{1}{16}$ sec = 16 cm/sec. Thus we shall see a circle of light where the linear velocity of the revolving disk $v \geqslant 16$ cm/sec. From the relationship between linear and angular velocity, $v = \omega R$, we have

$$R \geqslant \frac{16 \text{ cm/sec}}{\omega}; \quad \omega = 2\pi n = \frac{2\pi \times 30 \text{ rad}}{60 \text{ sec}};$$

$$R \geqslant \frac{16 \times 60}{2\pi \times 30} = 5 \cdot 1 \text{ cm},$$

i.e. the illuminated apertures merge into a continuous circle at a distance of $R \geqslant 5 \cdot 1$ cm from the centre.

16. We may consider the motion of a hoop rolling uniformly without slipping along a horizontal plane thus. Suppose that any two positions of the hoop be given. Then the hoop can be moved from one position to the other by moving it from the first position to the second in a straight line with a velocity equal to that of the centre, and by rotating the hoop uniformly about its centre with the same linear velocity so that all points on the circumference reach the places corresponding with its second position (Fig. 160). Since this is true for any two positions of the hoop, it is true for any two positions however close to-

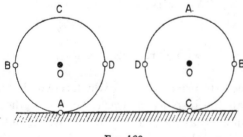

Fig. 160

gether. Therefore we can consider uniform rolling without slipping as a combination of two simultaneous motions: uniform
motion in a straight line with the velocity of the centre, and
uniform rotation about the centre with the same linear velocity
for the points on the circumference. But there is no acceleration
in uniform linear motion, while in motion in a circle all points on
the circumference have one and the same centripetal acceleration, which equals v^2/R.

17. When a cylinder rolls without slipping along a horizontal
plane, the line with which it makes contact with the plane at any
given moment is at rest and the cylinder itself is rotating about
this line. In the case given, then, the line of contact with the
board is clearly moving forward at a speed twice as great as that
of the axis of the drum. Therefore, when a man pushes the board
and moves forward a distance equal to the length of the board,
the drum will move forward a distance equal to half the length of
the board. Thus the man must move a distance of $2l$ in order to
reach the drum.

This can easily be verified with a round pencil and a ruler.

18. If the hoop is to roll along the plane without slipping, its
centre must move at the same velocity as the point of contact between hoop and plane. The linear velocity of this point, $v_1 = \omega R$.
Therefore if no slipping is to occur, v_1 must $= v$, or $\omega = v/R$.

19. When a wheel is rolling, it is turning, at every moment,
about its point of contact with the earth. Therefore the linear
velocities of its upper spokes are greater than those of its lower
ones, which are placed nearer the point which is stationary at
any given moment.

20. The aeroplane must start at mid-day and fly in the
direction opposite to the earth's rotation, i.e. from east to west,
at a speed equal to the earth's linear velocity on Leningrad's
latitude. The angular velocity of the earth

$$\omega = \frac{2\pi}{24} = \frac{\pi \text{ rad}}{12 \text{ hr}}.$$

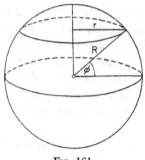

FIG. 161

The linear velocity on Leningrad's latitude (see Fig. 161) $v = r = \omega R \cos \psi = R\omega/2 = 6300 \times \pi/12 \times \tfrac{1}{2}$ km/hr $= 833$ km/hr. If he flies at this speed from east to west, the pilot will always see the sun in the same quarter as when he left Leningrad, i.e. in the south.

21. The first duellist must take into account that during the flight of his bullet, his opponent will have moved to another position. The time of flight of the first duellist's bullet does not depend on the roundabout's rotation: $t = R/v$. During time t, the roundabout will turn and point A will move to position B (Fig. 162a), i.e. through an arc of length $s = \omega R.t$; so he must

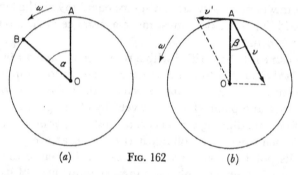

(a)   FIG. 162   (b)

fire in the direction OB, not in the direction OA. The angle α can be found from the equation $\alpha = \omega Rt/vt = \omega R/v$. The duellist standing on the circumference moves with a velocity of $v' = \omega R$, so the velocity of his bullet is compounded of two velocities v and v'. If he is to hit the centre of the roundabout, he too must not fire along AO, but at an angle which may be found from the formula

$$\sin \beta = \frac{\omega R}{v}$$

While
$$v' \ll v$$
$\beta \approx \omega R/v$, i.e. both duellists should aim at the same angle to the left of their respective opponents (if the rotation of the round-about is in the direction given in the diagram). But as ωR grows, β must increase, and the resultant velocity of the bullet will decrease and therefore the danger of being hit will decrease for the first duellist.

When $\omega R = v$, $\sin \beta$ must equal 1, i.e. the second duellist should aim in the direction opposite to v'. But in this case the resultant velocity of the bullet would be zero.

The time of flight of the second duellist's bullet t' depends on the roundabout's speed of rotation

$$t' = \frac{R}{v \cos \beta} = \frac{R}{v} \cdot \frac{1}{\sqrt{1 - \dfrac{\omega^2 R^2}{v^2}}} = \frac{1}{\sqrt{\dfrac{v^2}{R^2} - \omega^2}}$$

When $\omega R = v$, $t' = \infty$, i.e. the bullet of the second duellist 'hangs' in the air at A and he travels up to his own bullet (in which case, as we have seen above, the resultant velocity of the bullet equals zero).

When $\omega R > v$ the resultant velocity cannot anyhow be directed towards O, i.e. the first duellist cannot be hit by the second duellist's bullets. But the first can hit the second, if he selects angle α aright.

II. The Dynamics of Motion in a Straight Line

22. The plane flies horizontally with constant speed v. The bomb follows the path of a parabola, since its motion is compounded of horizontal motion with initial velocity v and uniformly accelerated vertical fall. If there were no air-resistance, the bomb's horizontal velocity would be no different from that of the plane, and the plane would be directly above the bomb the whole time—in particular, when the bomb hits the ground. But in fact, as a result of air-resistance, the bomb's horizontal velocity is decreasing all the time, and so it falls behind the plane (Fig. 163). Therefore the fall to earth and explosion of the bomb take place not underneath the plane, but considerably behind it.

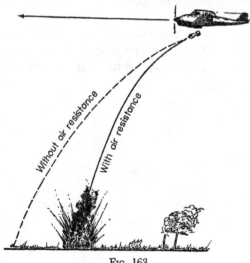

FIG. 163

23. The bullet's motion is compounded of two motions: (a) motion along a horizontal line at a constant velocity, which is the velocity at which the bullet left the barrel, and (b) falling due to the effect of the force of gravity, which begins at the moment the bullet leaves the barrel. As a result, the bullet will follow the path of a parabola. In a vertical direction the bullet will travel a distance $s = gt^2/2$, where t is the time during which it has been moving. The falling target will travel exactly the same distance vertically, since the force of gravity imparts the same acceleration to all falling bodies. Therefore the bullet will hit the target during its flight (case 1 in Fig. 164), or as the target hits the ground (case

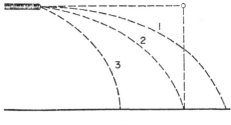

FIG. 164

2), given only that the distance from the marksman to the target is not farther than the bullet's range. If, however, this distance is greater than the bullet's range, then the bullet will not hit the target (case 3).

24. Two forces act on a falling drop: the force of gravity, which is constant and which accelerates the drop's motion, and the resistance of the air which retards the drop's motion and which grows as the drop's velocity grows. The force of air-resistance grows until it is equal to the force of gravity. Then the drop's velocity ceases to alter, and it continues to fall at a constant velocity.

If the dimensions of the drop are increased, the force of gravity is increased in proportion to the drop's volume, i.e. in proportion to the cube of the radius, while the force of air-resistance increases in proportion to the drop's cross-section, i.e. in proportion to the square of the radius. Therefore, as the drop's radius increases, the force of gravity increases faster than the air-resistance and therefore the constant speed at which the drop falls to earth also increases with the increase of the drop's dimensions.

25. Since the cross-section of both spheres is the same, they will both encounter the same air-resistance when travelling at the same velocity. But since the solid sphere is heavier than the hollow one, its acceleration will all the time be greater and it will accordingly fall faster.

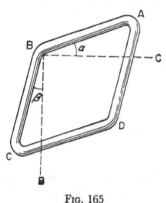

FIG. 165

26. When sides AB and DC of the rhombus (Fig. 165) are almost horizontal, it is at once obvious that the ball will roll faster down sides AD and DC (the second case in the problem). This can be seen from the fact that the ball will travel along side DC at a high average velocity, acquired from its motion along side AD. But in the first case the ball will travel along AB with a very small average velocity (since its acceleration is small).

The result found for this particular case remains true for the general case, as may be verified from the following calculation. Let sides AB and DC form an angle of α with the horizontal, and sides BC and AD an angle of β with the vertical. If the ball rolls along sides AB and BC it spends time $t_1 + t_2$ on this, where t_1 is spent on travelling along AB and t_2 on travelling along BC. The acceleration during motion along AB equals $g\sin\alpha$. Therefore to calculate t_1 we have the equation

$$A = \frac{g \sin \alpha \, t_1^2}{2}.$$

The acceleration during motion along BC equals $g \cos \beta$, and this motion takes place with an initial velocity of $\sqrt{2Ag \sin \alpha}$, so we can find t_2 by solving the equation

$$A = \sqrt{2Ag \sin \alpha}\, t_2 + g \cos \beta\, \frac{t_2^2}{2}.$$

For the second case we shall have the same expressions with the only difference that the accelerations $g \sin \alpha$ and $g \cos \beta$ must change places. So the sum of the two times for the first case will be

$$\sqrt{\frac{2A}{g \sin \alpha}} + \frac{-\sqrt{2Ag \sin \alpha} + \sqrt{2Ag \sin \alpha + 2Ag \cos \beta}}{g \cos \beta},$$

and for the second case the sum of the two times will be

$$\sqrt{\frac{2A}{g \cos \beta}} + \frac{-\sqrt{2Ag \cos \beta} + \sqrt{2Ag \cos \beta + 2Ag \sin \alpha}}{g \sin \alpha}.$$

Since it is clear that $g \sin \alpha$ is less than $g \cos \beta$, we find that the sum of the two times in the first case is greater than in the second.

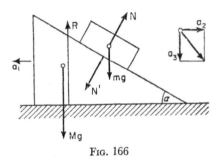

FIG. 166

27. Let us consider the forces acting on the load m and on the wedge M (Fig. 166). The load m is subject to: (1) its weight mg and (2) the reaction of the wedge N. The wedge is subject to (1) its own weight Mg, (2) the pressure exerted by the load N' and (3) the reaction of the plane, R. As a result of the horizontal component of the pressure exerted by the load, the wedge moves to the left relative to the plane with a horizontal acceleration a_1, which can be found from the equation

$$Ma_1 = N' \sin \alpha. \qquad (1)$$

In the vertical direction the wedge has no acceleration, therefore

$$Mg - R + N' \cos \alpha = 0 \qquad (2)$$

Let us call the horizontal component of the load's acceleration *relative to the wedge* a_2, and the vertical component a_3. Then the horizontal component of the load's acceleration *relative to the plane* will be a_2-a_1, and the vertical component will be a_3. These accelerations may be found from the equations

$$m(a_2-a_1) = N \sin \alpha \tag{3}$$

and

$$ma_3 = mg - N \cos \alpha \tag{4}$$

Plainly $N' = N$ and

$$a_3 = a_2 \tan \alpha. \tag{5}$$

From equations (4), (5), (3) and (1) we find that the pressure of the load on the wedge is

$$N = \frac{mMg \cos \alpha}{M + m \sin^2 \alpha} \tag{6}$$

Now from equation (2) we can find the pressure of the wedge on the plane

$$R = \frac{Mg(1 + m \cos^2 \alpha)}{M + m \sin^2 \alpha}.$$

Further from (1) and (6) we find the wedge's acceleration

$$a_1 = \frac{mg \cos \alpha \sin \alpha}{M + m \sin^2 \alpha}. \tag{7}$$

From (3), (6) and (7) we find the horizontal component of the load's acceleration relative to the wedge

$$a_2 = \frac{(M + m)\, g \cos \alpha \sin \alpha}{M + m \sin^2 \alpha}, \tag{8}$$

and the horizontal component of the load's acceleration relative to the plane

$$a_2-a_1 = \frac{Mg \cos \alpha \sin \alpha}{M + m \sin^2 \alpha}.$$

From (8) and (5) we find that the vertical component of the acceleration of the load relative to the plane

$$a_3 = \frac{(M + m)\, g \sin^2 \alpha}{M + m \sin^2 \alpha}.$$

28. Let us resolve force 5 into two components acting along the sides of the square (Fig. 167). The forces obtained, 1′ and 2′, will each equal 1 kg. Therefore forces 2 and 2′ will balance each other and a force of 2 kg will act at point A along AD. Let us then shift force 6 along its line of action to D and resolve it into forces acting along the sides of the square. We shall get forces 3′ and 4′, which are each of 1 kg. Therefore forces 4 and 4′ balance each other and a force of 3 kg will act in the direction AD. It remains to

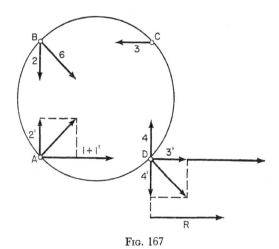

Fig. 167

compound this force with force 3, which acts in a direction parallel but opposite to that of the 3 kg force. The resultant R, of magnitude 2 kg, will plainly act in a direction parallel to its components along a straight line which lies at a distance equal to half the side of the inscribed square from AD. The ring will move, as a result of this force, in a straight line under acceleration, and will at the same time rotate (since the force has a moment relative to the centre of the ring) also under acceleration. This conclusion could have been reached in general terms at the beginning by considering that forces 1, 2, 3, 4 constitute two couples which rotate the ring in one direction, while forces 5 and 6 have a resultant which will cause the ring to move in a straight line under acceleration.

29. When the scale-pan with the weight on it is at rest, the spring is extended to such an extent that the spring's elasticity balances the weight of scale-pan plus weight. If the spring is forced out of this position and pulled downwards a little, when it is released it will make oscillations about the position of equilibrium. The spring's extension will then be least when the scale-pan is in its highest position. For the weight to cease to exert pressure on the scale-pan, it is necessary that the pan should begin to move downwards from its highest position with the same acceleration as the weight, i.e. with an acceleration equal to that imparted by the force of gravity, g. This will obviously be the case when the spring is not stretched at all in its highest position. Since the highest position is symmetrically placed to the lowest position relative to the position of equilibrium of weight plus pan, it follows that for the scale-pan to reach a position in which the spring is not extended at all, the spring must first be extended through a distance equal to that through which the scale-pan and weight have stretched it in reaching their position of equilibrium.

Consequently the scale-pan and weight must be pulled downwards with a force equal to their combined weight.

30. The spring is compressed initially (by comparison with its normal length) by the weight of the upper lamina by an amount

$$x_1 = \frac{m_1 g}{k}.$$

For the spring to raise the lower lamina on expansion, it must be extended, by comparison with its normal length, by an amount greater than

$$x_2 = \frac{m_2 g}{k}.$$

Consequently, the upper lamina must be depressed with such a force that when it is released it should jump up through a height greater than

$$x_1 + x_2 = \frac{(m_1 + m_2)g}{k},$$

reckoning from the position which the upper lamina occupied before it was depressed. But the upper lamina will jump up on release through the same distance above the position of equilibrium as it lies below this position when depressed. Thus, if the upper lamina is to carry the lower one up with it a little, it must be depressed through a distance greater than $x_1 + x_2$. For this, the upper lamina must be depressed with a force greater than

$$k(x_1 + x_2) = (m_1 + m_2)g.$$

31. Since the cyclist is moving uniformly, all the forces acting on him must balance each other. Thus, the reaction of the plane must be equal and opposite to the weight of the cyclist and his bicycle, i.e. it must act vertically upwards. Thus the re-action of the plane in the instance we are considering is not perpendicular to the in-clined plane (Fig. 168) and must have, besides its normal component N', a component T, acting along the plane. This component is the force of fric-

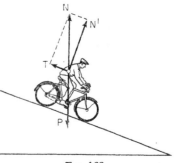

Fig. 168

tion which is exerted on the bicycle by the plane. If there were no force of friction between plane and bicycle, the cyclist would move down the plane with a constant acceleration and not uniformly, in spite of his brakes.

32. For the plank not to slip, the man's legs should exert a force along it which balances the component of the force of gravity of the plank, i.e. which equals $Mg \sin \alpha$, and which is directed upwards. The plank will then exert an equal and opposite force on the man, i.e. $Mg \sin \alpha$, acting downwards along the plank. Besides this, the man is subject to the component of his own weight $mg \sin \alpha$. Consequently the man must move down the length of the plank with an acceleration relative to the earth such that

$$ma = Mg \sin \alpha + mg \sin \alpha,$$

hence

$$a = g \sin \alpha \left(1 + \frac{M}{m}\right).$$

33. If a body is falling freely and during the time of its fall a change takes place in the relative positions of its component parts within the body, the body's centre of gravity continues to move with the acceleration of free fall, since displacements which take place as a result of the effect of internal forces cannot alter the position of the centre of gravity. Therefore the centre of gravity of the system test-tube-plus-fly will also move as before with the acceleration of free fall. Consequently, when the fly flies up from the lower part of the test-tube to the upper part, the bottom of the test-tube will be lowered somewhat relative to the centre of gravity of the whole system (test-tube-plus-fly). Thus the bottom of the test-tube will strike the ground sooner than if the fly were to remain motionless. In other words, the length of time of the fall is reduced.

34. Equilibrium will not be destroyed if the bird takes off and hovers in the air. The explanation of this is that when the bird hovers, it must force the air downwards in order to create an up-thrust which can support it in the air. The air thus forced downwards will create an additional pressure on the bottom of the box and the average magnitude of this pressure will exactly equal the bird's weight. It is true that when the bird takes off and when-ever the bird makes any sudden movements (when it will move with acceleration), the magnitude of this pressure may alter, so that the scales begin to swing about the equilibrium position, but on average equilibrium will not be destroyed.

35. If a balloon descends with constant velocity, the resist-ance of the air is directed upwards, and therefore the balloon's weight P is balanced by the upthrust Q plus the air-resistance F_1

$$P = Q + F_1. \tag{1}$$

If, after ballast of weight X has been jettisoned, the balloon begins

to rise with the same velocity v, the force of air-resistance will change its direction, i.e. it will act downwards. Therefore we shall get:
$$P - X = Q - F_2. \tag{2}$$
The magnitude of the air-resistance is proportional to the velocity; since the velocities of descent and ascent are the same, the air-resistance is the same in both cases $(F_1 = F_2)$. On the basis of this, we find from equations (1) and (2):
$$X = 2(P-Q) = 2F_1.$$
36. At any point of its path upwards the bullet's full acceleration is directed downwards and equals
$$g + \frac{f}{m},$$
and at any point of its path downwards, it is directed downwards and equals
$$g - \frac{f}{m},$$
where m is the mass of the bullet and f the force of air-resistance, which always acts in a direction opposite to that of motion. The force f of the air's resistance increases with the body's velocity. But the bullet will have maximum velocity at the very lowest point of its flight—at the moment of firing and, on the way back, at its moment of fall. (Notice, by the way, that the maximum velocity on its return path is less than the maximum velocity on the way up since part of the bullet's energy is expended on over-coming the air-resistance.)

Consequently, the force of air-resistance f is first of all directed downwards and will be greatest at the lowest point (at the moment of firing), then it will decrease as the velocity decreases, it will pass through zero (at the highest point of the flight), and, changing its direction for the return to ground, will again begin to increase. The bullet's acceleration will be directed downwards throughout however and will be greatest at the moment of firing, then it will decrease and will reach its lowest value at the moment of fall.

37. So long as the tube is stationary, the sphere lies in the tube and the elastic force F of the compressed spring acts on it. Therefore, when the tube begins to fall freely, the sphere first of all falls with less acceleration than the tube and lags behind it; the spring expands. As a result of the effect of the spring's elasticity, the sphere not only lags behind the tube in a vertical direction, but also acquires velocity in a horizontal direction. Therefore it will appear at the mouth of the tube with a certain horizontal velocity,

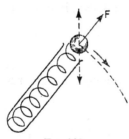

FIG. 169

and then, like every body thrown with an initial horizontal velocity, the sphere will fall along the path of a parabola (Fig. 169).

38. The trolley's acceleration will be reflected in the fact that the weight will be inclined at an angle and the fact that the thread by which the weight is suspended will be subject to an increased tension. But the moment acting on the scales' beam is determined by the component of the force which acts perpendicularly to the arm, i.e. the vertical component of the thread's tension, which will not be changed. This may be seen from the fact that the load, while inclining, remains stationary relative to the trolley, i.e. its acceleration in a vertical direction equals zero, and therefore the vertical component of the thread's tension balances the weight of the weight. But if the moment which acts on the left-hand arm of the beam does not alter, then the tension in the spring will also remain unaltered.

39. In the first case acceleration is imparted to the piston by its own weight and by the pressure of the compressed air in the vessel (since the volume of the cylindrical part is small by comparison with the vessel's volume, we may consider that the air's pressure is constant throughout the period of the piston's motion in the cylinder). But the sphere will not exert pressure on the piston, since it falls only under the force of gravity, i.e. with less acceleration than the piston, from which it is therefore separated.

The sphere does not therefore alter the piston's acceleration and so does not alter the time during which it moves. In the second case the pressure exerted by the gas is the same, but the piston's mass is greater. So the piston's acceleration will be less, and the time during which the piston moves in the cylinder will be greater than in the first case.

40. If A pulls as before, the dynamometer will register 42 kg, but then A will impart to B an acceleration towards himself and B will lean forward.

41. For the coupling to be undone, it must not be taut. Consequently, this is possible only when the train is going downhill or else along level ground with the brakes applied. But then, after the carriages are uncoupled, they can only fall back from the rest of the train when the slope comes to an end or the brakes are taken off (or else the uncoupled carriages must be subjected to braking).

42. Let us suppose, for clarity, that m_1 $> m_2$ (Fig. 170). Then m_1 will descend and m_2 will rise. Since we are neglecting the mass of the pulley, the tension in both parts of the cord must be the same ($T_1 = T_2$). In fact, if the pulley has no mass, no moment is required to impart to it angular velocity. Therefore the moments of both the forces of tension in the two parts of the cord are equal, and since they act equidistantly from the pulley's centre, the tensions themselves are

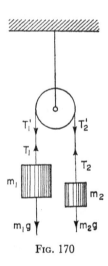

FIG. 170

also equal, i.e. $T_1' = T_2'$. The motion downwards of load m_1 takes place as a result of the effect of two forces, m_1g and T_1, which act in the directions shown in Fig. 170. Therefore, if the acceleration of load m_1 is a, we shall have:

$$m_1a = m_1g - T_1. \qquad (1)$$

Load m_2 moves upwards with the same acceleration a, under the effect of two forces T_2 and m_2g, so that

$$m_2a = T_2 - m_2g. \qquad (2)$$

From equations (1) and (2) we get

$$a = \frac{(m_1 - m_2)\,g}{m_1 + m_2}, \qquad T_1 = T_2 = \frac{2m_1m_2g}{m_1 + m_2},$$

The force exerted on the axis of the pulley equals the sum of the tensions in the two parts of the cord, i.e.

$$P = T_1 + T_2 = \frac{4m_1m_2g}{m_1 + m_2}.$$

43. Let us consider both loads as points of matter lying at the loads' respective centres of gravity. Load M_2 begins to descend with a linear acceleration a, which is not equal to g, since the presence of the load M_1 causes the rod to exert an upward force F on M_2. Therefore, from the second law of Newton, we have

$$M_2a = M_2g - F \qquad (1)$$

Load M_1 begins to rise with the same acceleration a, since the presence of load M_2 causes the rod to exert an upward force of F' on M_1. Therefore

$$M_1a = -M_1g + F' \qquad (2)$$

Since we can neglect the mass of the rod, the moments of the forces F and F', acting on the rod, relative to the axis of rotation, must be equal and opposite and their sum must accordingly be zero. (Otherwise a rod of infinitely small mass would be given infinitely great angular acceleration.) But the distance of these forces from the fulcrum is the same for both, therefore $F = F'$.

Then from equations (1) and (2) we get:

$$a = \frac{(M_2 - M_1)g}{M_1 + M_2}; \quad F = \frac{2M_1M_2g}{M_1 + M_2}.$$

The resultant of forces F and F' is equal and opposite to the force of the rod's pressure on its axis at the initial moment. Consequently this last force equals

$$R = \frac{4M_1M_2}{M_1 + M_2}\, g = 3\cdot5 \text{ kg.}$$

The problem we have just considered is in principle no different from problem 42. In this one the role of the pulley was played by the rod, fixed at its mid-point.

44. Let the mass of the whole chain be m, and the mass of unit length of the chain be $m_0 = m/l$. Let a part of the chain of length x be hanging down from the table at any moment; to begin with, $x = l/2$, where l is the length of the whole chain. Then the force which causes the chain to move will be proportional to the length of chain hanging down, i.e. $m_0 xg$. The acceleration of the chain will equal $m_0 xg/m$. To begin with, when $m_0 x = m/2$ the acceleration equals $g/2$, but then it increases so that the chain's motion is not uniformly accelerated.

If identical masses M be attached to the ends of the chain, then at the moment when a part of the chain of length x is hanging from the table, the value of the moving force will be $m_0 xg + Mg = (m_0 x + M)g$, and the chain's acceleration at that moment will be $(m_0 x + M)x/(m + 2M)$. To solve the problem as to which case will cause the chain to slip off faster, we must find in which case will the acceleration increase at the greater rate. For this we must compare the two expressions:

$$\frac{m_0 xg}{m} \quad \text{and} \quad \frac{(m_0 x + M)g}{m + 2M}.$$

To compare these fractions, we shall put them over a common denominator and compare their numerators. The numerator of the first fraction will be $m_0 xgm + 2m_0 xgM$, and that of the second fraction will be $m_0 xgm + Mgm$.

It is clear that these numerators are equal at the beginning when $m_0 x = m/2$. In the subsequent moments, the acceleration will be greater in the first case than in the second.

So the chain will slip off more quickly when there are no masses attached to its ends.

This result can also be obtained from the law of conservation of energy, taking into account that at the moment when the end of the chain slips off the table, its centre of gravity is at a distance of

$l/2$ below the edge of the table. Then if v_1 be the velocity at that moment and if there are no masses attached to the end,

$$mg\frac{l}{2} = \frac{mv_1^2}{2}$$

(Potential Energy) = (Kinetic Energy)

or $gl = v_1^2,$

while if masses are attached to the ends

$$(m + M)g\frac{l}{2} = \frac{m + 2M}{2}v_2^2$$

or

$$gl = \frac{(m + 2M)v_2^2}{m + M},$$

i.e. the time of slipping for the chain must be less when there are no masses attached to the ends.

45. Since the pulleys are weightless, $T_2 = T_3$ and $T_1 = 2T_2$

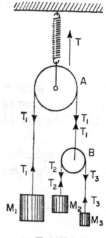

FIG. 171

(see problem 42). Also $T = 2T_1$. Load M_1 moves downwards with an acceleration a, under the action of two forces: M_1g (acting downwards) and T_1 (acting upwards) (Fig. 171). Therefore

$$M_1a = M_1g - T_1. \tag{1}$$

If pulley B were not rising with an acceleration a, load M_2 (for clarity, we shall take $M_2 > M_3$) would move downwards with acceleration a' and load M_3 would move upwards with the same acceleration. Taking the movement of pulley B upwards into account we shall have the following equations for the movement of loads M_2 and M_3:

$$M_2(a' - a) = M_2g - T_2, \tag{2}$$
$$M_3(a' + a) = -M_3g + T_2. \tag{3}$$

From (2) and (3) we have that

$$a' = \frac{M_2 - M_3}{M_2 + M_3}(a + g).$$

Substituting this value for a' in equation (2), we shall obtain:

$$T_2 = \frac{2M_2 M_3}{M_2 + M_3}(g + a) = \tfrac{1}{2}T_1.$$

Now from equation (1), we shall find that

$$a = \frac{M_1 M_2 + M_1 M_3 - 4M_2 M_3}{M_1 M_2 + M_1 M_3 + 4M_2 M_3}g$$

and

$$T = 2T_1 = \frac{16 M_1 M_2 M_3}{M_1 M_2 + M_1 M_3 + 4M_2 M_3}g.$$

46. Two forces are acting on the chain: the force of gravity of the hanging part and the force of friction between the table and the part of the chain lying on the table. When length l_1 of chain hangs down the force of static friction balancing the weight of this length has its maximum value. Let the mass of unit length of the chain be m_0 $(m_0 = m/l)$ and the coefficient of static friction be k, i.e. the ratio of the greatest value for the force of static friction to the normal pressure. Then the fact that the forces are equal

$$m_0 g l_1 - k m_0 g (l - l_1) = 0$$

gives us that

$$k = \frac{l_1}{l - l_1}.$$

47. A train usually stops in such a position that the couplings between trucks are taut (since the buffers, which are compressed during braking, push the trucks apart and tauten the couplings). Therefore the locomotive, if it is to move the train forward, must overcome the static friction of all the trucks at once. But if it pushes the train backwards, first it pushes only the first truck and in doing this it has to overcome the static friction of one truck. Once the truck has started moving static friction gives place to the sliding friction of the ball-bearings in the wheel-axles and the rolling friction of the wheels on the rails. These forces are considerably less than that of static friction. Therefore when it pushes the first truck and the second truck backwards, the locomotive is able to overcome the second truck's static friction and so on. After the whole train has started to move backwards and the buffers of all the trucks are compressed, the locomotive changes to forward gear. Then, as it tautens the coupling of the first truck, it is overcoming at first the static friction of only the first truck, then that of only the second, and so on, until the whole train is moving. The compressed buffers between the trucks also help this.

48. The braking of a car, bus or tram consists in the creation of a moment by the brakes, which causes retardation of the wheels' rotation. But if the wheels' rotation were retarded without a reduction in the vehicle's speed, the wheels would begin to slip on the ground in a forward direction. When this happens the ground exerts a force of friction which hinders slipping from taking place, i.e. it acts in a backward direction. This force reduces the vehicle's speed and finally brings it to a stop. Therefore the ground is the outside body which exerts a force that changes the vehicle's velocity.

49. Suppose that the head lies at the left end of the broom-handle. The centre of gravity of handle plus head lies between the supporting hands and nearer to the left forefinger. The pressure exerted by the whole broom will then not be equal on both hands: it will be greater for the left hand than for the right. Therefore the force of friction between handle and finger, which is proportional to the pressure, will be greater for the left hand also. If the right hand be moved towards the left, the broom will remain still, since the greater force of friction between it and the left finger will hold it in place, until both fingers are at equal distances from the broom's centre of gravity. Then as the right finger moves farther the pressure will become greater on it than on the left finger, and the broom will move with the right finger until the centre of gravity again lies halfway between the two fingers. So it will continue until the two fingers come together.

If the left hand be moved towards the right hand, the broom will move with the left finger, sliding over the right finger until both fingers are at the same distance from the centre of gravity. The broom's farther motion will be as in the preceding case.

If the hands be moved towards one another, the broom will move with the left hand until the broom's centre of gravity is equidistant from both fingers. Then both fingers will slide under the broom, which will remain stationary. Since the centre of gravity lies between the fingers in all these cases, the broom remains in equilibrium and does not fall.

50. When a car's brakes are applied the ground begins to exert a force of friction F on its wheels (for the sake of simplicity, we shall assume that it exerts this force only on its rear wheels). To consider how this force affects the motion of the car's centre of gravity, let us imagine that two forces F' and F'' are applied to the centre of gravity; each of these forces is equal to F in magnitude and their line of action is parallel to that of F, but they act in opposite directions to each other (Fig. 172). We can do this, since the application of two

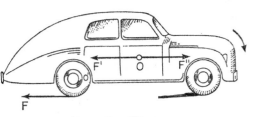

FIG. 172

equal and opposite forces will not alter the car's motion. But these three forces can be considered as a force F' and a couple consisting of F'' and F. It is not hard to see that force F' will brake the car, while the couple F'' and F will cause rotation in the direction indicated by the arrow, i.e. it will cause the car's bonnet to dip.

If the brakes also act on the front wheels and the force of

friction accordingly acts between the ground and both pairs of wheels, the result will be the same, since if a force of friction F be given, its moment relative to O does not depend on how this force is distributed between the wheels.

51. The velocity of the body at B depends on how much of its potential energy the body will expend on work done against the force of friction. The forces of friction on routes 1 and 2 are not the same, since the pressures exerted by the body on the surface are not the same. This is conditioned by the following factors.

When moving along any curved path a body must have a centripetal acceleration. On route 1 this acceleration is directed downwards, on route 2 upwards.

This acceleration, whose direction is normal to the plane along which the body slides, arises because the component of the force of gravity which is normal to the surface and the normal reaction of the surface on the body are not equal in magnitude. The resultant of these two forces (their difference) is what imparts to the body the necessary centripetal acceleration. Therefore on route 2, where the centripetal acceleration is directed upwards, the normal reaction is everywhere greater than the normal component of the force of gravity, and on route 1 it is less everywhere. The magnitude of the normal components of the force of gravity are in general different for routes 1 and 2, since the angles between the normal to the plane and the vertical are different. But since both surfaces are of the same radius, the normal components of the force of gravity alter within the same limits on both routes, but in reverse order. Therefore we can say that the normal reaction of the surface on the body (and so of the body on the plane) is on average less on route 1 than on route 2. Since the coefficient of friction on both routes is the same, the force of friction is less on route 1 than on route 2, i.e. the work which is expended on overcoming the force of friction will be less on route 1. Thus at point B the body has greater kinetic energy, and so greater velocity, if it moves along route 1, not along route 2.

52. Equilibrium will be broken since the tension in the rope from which the swinging load is suspended cannot remain constant and equal to the weight of its load. At the limits of the swing, where the load's velocity is zero, the tension in the rope will be equal to the component of the load's weight acting in the line of the rope, i.e. it will be less than the load's weight. At the centre point of the swing the tension in the rope must not only balance the load's weight, but must also impart to it the necessary centripetal force upwards, i.e. its tension must be greater than the load's weight. Therefore if the right-hand load swings, the left-hand load can not remain at rest and the resulting motion is of a more complicated character than that of a pendulum's oscillations, suspended from a fixed point. Experiment and calculation show that the swinging load will outpull the other.

53. If the driver applies his brakes, the car will stop when its kinetic energy has been expended in work against the force of friction. On the other hand, when the car turns the same force of friction will play the part of a centripetal force which makes the car move along the arc of a circle.

In the case in which the brakes are applied

$$\frac{mv^2}{2} = Fx,$$

where F is the force of friction, and x is the distance which the car will travel after the brakes are first applied. Hence $x = mv^2/2F$. Evidently, if the car is not to be smashed up, we must have

$$x = \frac{mv^2}{2F} \leqslant a$$

or

$$F \geqslant \frac{mv^2}{2a}$$

In the case where the car is turned

$$F = \frac{mv^2}{R}.$$

and if the car is not to be smashed up, we must have

$$R = \frac{mv^2}{F} \leqslant a$$

or

$$F \geqslant \frac{mv^2}{a}.$$

To avoid crashing into the wall, braking requires half the force of friction which a turn requires. Clearly it is better to brake rather than to turn.

III. Statics

54. No. A man can raise his left leg and not lose balance only in the case when the vertical line passing through his centre of gravity passes also through the sole of his right foot. In the position described, this cannot be so.

55. 1st case: all three equal forces act in the same plane and make an angle of 120° with each other, their resultant equalling zero (Fig. 173a). 2nd case: the pike and the crab pull in directly opposite directions while the swan pulls vertically upwards (Fig. 173b); then the swan's strength would be less than the weight of the cart (though the last condition is not given in the fable).

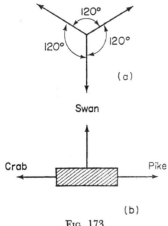

Fig. 173

56. This is done to make the 'arm' of the application of the force as large as possible; the larger the 'arm', the less force needs to be applied in order to overcome the effect of the tight spring which raises the trolley-bus's arm and presses it to the wire (Fig. 174).

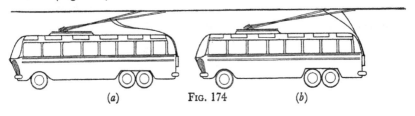

(a)   Fig. 174   (b)

57. Let us resolve the weight of lamp and shade in two directions at right angles to one another: along the arm *AB* and along the line of *AE*, the height of the isosceles triangle *ACD*, produced (Fig. 175a). Since arm *AB* makes an angle of 30° with the line of hang of the lamp, we can easily find that the force which acts in the line of *AE*, the height of the triangle *ADC*, and which exerts a stress on the upper arm *AC* and *AD*, equals $\frac{1}{2}$ kg, while the force which exerts a pressure on arm *AB* equals $\sqrt{1 - (\frac{1}{2})^2} = \frac{\sqrt{3}}{2}$ kg.

Now let us resolve the $\frac{1}{2}$ kg force along arms *AC* and *AD* (Fig. 175b). In the case given, the parallelogram of forces will be a

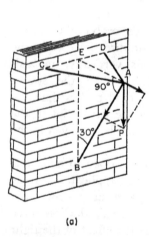

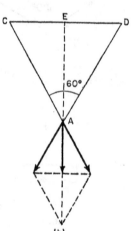

(a)   Fig. 175   (b)

rhombus, in which, of course, the diagonals are at right angles to each other, and since the angle at vertex *A* of the rhombus equals 60°, we can easily find that the unknown force *F*, which exerts a stress on each of the upper arms satisfies the equation

$$F^2 - \left(\frac{F}{2}\right)^2 = (\tfrac{1}{4})^2$$

hence

$$F = \frac{1}{2\sqrt{3}} \quad .$$

58. The strings will break more easily, the nearer they are to a horizontal position. To explain this, let us consider the tension in each string. Tensions of F_1 and F_2 act along the strings. These two forces and the weight P of the body are in equilibrium, i.e. the resultant F of forces F_1 and F_2 is equal and opposite to force P (Fig. 176). If the angle be-

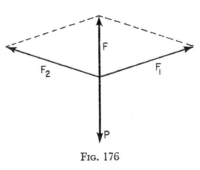

tween F_1 and F_2 is 120°, the triangle will be equilateral, and the tension in each string equals the weight of the load. If the angle is less than 120°, each tension is less than the weight of the load and de- creases as F_1 and F_2 come closer to P's line of action. On the other hand, the nearer F_1

FIG. 176

and F_2 move to the horizontal, the greater their value. When the strings adopt an almost horizontal position, the tension in them is very great and they break easily.

59. The solution to this problem is analogous to that of the preceding one. If the rope is already taut, a force P, acting at right angles to the rope, can create great tension in it. This may be seen from the fact that force P must equal the sum of the ten- sions F in the two parts of the rope which meet at the point of application of P (Fig. 177). Then if angle α is close to 180°, the tension F will be many times greater than the force P with which the man pulls the rope.

60. Let us transfer force P which acts on wedge I to A and

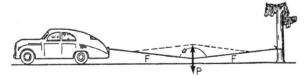

FIG. 177

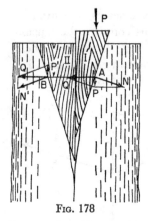

FIG. 178

resolve it at that point into a force N, perpendicular to the surface of the wedge, and a force Q, parallel to the base of the wedge (Fig. 178). Force N produces an equal and opposite reaction in the log. Force Q acts on wedge II, which was already inserted, and produces in it a reaction. Let us transfer force Q to point B, where wedge II is in contact with the log. Let us resolve force Q into force N', perpendicular to the wedge's surface, and force P', perpendicular to the base of the wedge. Force N' produces an equal reaction in the log, force P' drives out wedge II.

61. On the first bar the weights are of volumes proportional to their respective weights. When they are lowered into the water, the forces acting on the arms of the bar will be altered (according to Archimedes' principle) by amounts which are proportional to their volumes. Therefore the relative change of forces acting on the bar will be the same for both ends. Consequently equilibrium will not be destroyed under water.

On the second bar, both weights will 'lose weight' under water by the same amount and since their weights are different, the relative 'loss of weight' will be greater for the one which weighs less. The ratio between the forces acting on the arms of the bar will alter and equilibrium will be broken. The end of the bar at which the relative loss of weight is less will tilt downwards, i.e. the end with the 3 kg weight.

62. Generally speaking, no, since at every point on the earth's surface there is a magnetic inclination (the lines of force of the earth's magnetic field are not horizontal).

63. The moment of the force of gravity for the larger load will always be greater than the moment of the force of gravity for the lesser load, since the perpendicular distances of the directions of

these forces P_1 and P_2 from the fulcrum remain equal whatever the
position of the beam (Fig. 179).
But if the beam moves, the centre
of gravity of the beam and pointer
is changed: it moves from O to
O', and there arises the additional moment of the force of

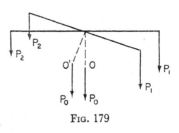

Fig. 179

gravity for the beam and pointer. This moment is in addition to
the moment of force P_2 and the position of equilibrium is determined by the equality of moments

$$P_1 = P_2 + P_0.$$

64. When the sections AB, CD and EF of the rope are parallel
and the system is in equilibrium (Fig. 180), the tension in the
rope must equal half the weight of load M_1, and load M_2 must

be half M_1. If the rope's
point of attachment A is
then shifted to the right
to A', the relevant sections of the rope are no
longer parallel and a tension in them which was
equal to half the weight
of M_1 would no longer be
able to support the pulley
from which M_1 hangs.
Consequently equili-

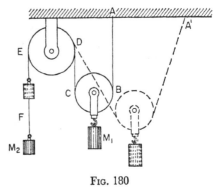

Fig. 180

brium is destroyed, load M_1 descends and load M_2 rises.

65. Let the force with which the man pulls rope a be x kg (Fig.
181). Then the tension in rope b will also be x kg. The tension in
rope c is balanced by the combined effect of two parallel forces
x and x, so that it equals $2x$. This must also be the tension in
rope d, which is a continuation of rope c. The plank hangs from
two ropes b and d (rope a is not fastened to the plank and therefore does not support it). The tension in b equals x kg, that of d

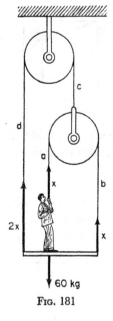

FIG. 181

equals $2x$ kg; the sum of these parallel forces is $3x$ kg, which acts upwards. On the other hand a force acts downwards on the plank, that exerted by the man. The man weighs 60 kg, but rope a is pulling him upwards with a force of x kg, so the force of the man on the plank is $(60-x)$ kg. The sum of all the forces acting on the plank must be equal to zero, since the plank is in equilibrium, i.e. $(60-x)-3x = 0$, from which $x = 15$ kg.

66. Four forces are acting on pulley B (Fig. 182): force P, force F and the tensions T in the two sections of the rope either side of the pulley (these tensions are equal). Since the pulley is in equilibrium, the sum of the resolutes of all the forces in any direction equals zero. Resolving horizontally: $F-T\cos 60° = 0$; resolving vertically; $P-T-T\cos 30° = 0$. From these two equations we obtain that $F = P(2-\sqrt{3})$.

67. Imagine that not one hole but two are cut in the lamina, both the same, as shown in Fig. 183. Then the centre of gravity of the lamina will lie at the centre of the circle O, and its weight will be $\frac{1}{2}P$, where P is the weight of the solid lamina (since the

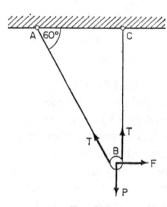

FIG. 182

FIG. 183

weight of each of the parts removed is equal to $\frac{1}{4}P$). Now fill in the left-hand hole. This will add a weight equal to $P/4$, applied at the centre of the filled in hole, i.e. at C', which lies at a distance of $R/2$ from the centre of the lamina. Then the point of application of the resultant of these two forces, i.e. the centre of gravity, lies between C' and O. Let the distance of this point from O be x. Then

$$\frac{P}{2}x = \frac{P}{4}\left(\frac{R}{2} - x\right),$$

hence

$$x = \frac{R}{6}.$$

This same result can be reached in a different way. Fill in the hole with a little circle and apply a force at its centre C equal to the weight of this circle, i.e. $\frac{1}{4}P$, and acting vertically upwards. Then we shall have two parallel and opposite forces: the weight of the whole lamina P, acting downwards and applied at O, and the force of $\frac{1}{4}P$, acting upwards and applied at C. The point of application of their resultant, i.e. the centre of gravity, lies on the line OC to the left of O. Let x be its distance from O. Then

$$Px = \frac{P}{4}\left(x + \frac{R}{2}\right) \text{ and } x = \frac{R}{6}.$$

68. Let us first find the position of the centre of gravity of the first two spheres. Clearly this point divides the distance x_2-x_1 into lengths which are in inverse proportion to the masses M_1 and M_2. Let us call its distance from the end of the bar x (Fig. 184). Then we shall have:

$$\frac{x - x_1}{x_2 - x} = \frac{M_2}{M_1}.$$

Hence we shall obtain:

$$x = \frac{(M_1 x_1 + M_2 x_2)}{M_1 + M_2}.$$

At this point acts the sum of the weights of the first two spheres, which we shall call M. Then if we apply the results we have ob-

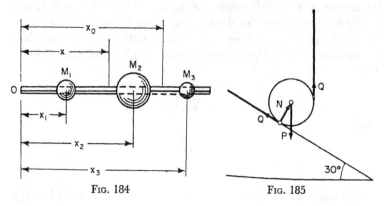

FIG. 184   FIG. 185

tained to the masses M and M_3, which are at distances from the end of the bar of x and x_3 respectively, we shall find that the centre of gravity of the system of all three spheres lies at a distance of

$$x_0 = \frac{Mx + M_3 x_3}{M + M_3}$$

from the end of the bar; substituting for x and M, we shall obtain

$$x_0 = \frac{M_1 x_1 + M_2 x_2 + M_3 x_3}{M_1 + M_2 + M_3}$$

or in general for any number of spheres

$$x_0 = \frac{\sum_{1}^{n} M_i x_i}{\sum_{1}^{n} M_i}.$$

69. Since the cylinder is in equilibrium, the sum of the resolutes in any direction of all the forces acting on the cylinder must equal zero. Resolving all the forces vertically (Fig. 185), we shall obtain the equation

$$Q + \frac{Q}{2} + N\frac{\sqrt{3}}{2} - P = 0.$$

Resolving all the forces horizontally, we shall obtain the equation

$$\frac{N}{2} - Q\frac{\sqrt{3}}{2} = 0.$$

From these two equations we can easily find that

$$Q = \frac{P}{3}.$$

70. To rise on to the step, the wheel must rotate about point A (Fig. 186). For this the moment of force F about A must be greater than or equal to the moment of the force of gravity for the wheel about the same point.

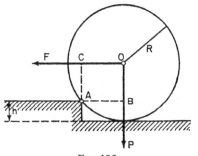

If we drop perpendiculars from A on to the line of action of the force of gravity and on to that of force F, we shall find the distances

Fig. 186

from A of the lines of action of these forces, AB and AC. For the wheel to roll up on to the step, the inequality must hold good:

$$F.AC > mg.AB$$

or

$$F > mg\frac{AB}{AC}.$$

Since $AC = R-h$,

$$AB = \sqrt{R^2 - (R-h)^2},$$

and finally

$$F > mg\frac{\sqrt{R^2 - (R-h)^2}}{R-h} = mg\frac{\sqrt{h(2R-h)}}{R-h}$$

and if $h \ll R$, then

$$F \geqslant mg \sqrt{\frac{2h}{R}}.$$

71. There is equilibrium in both cases when the component of the weight P of load m_2 along the line of the rope equals the tension T in the rope, i.e. the weight of load m_1 (Fig. 187). To find whether this state of equilibrium is stable or not, we must con-

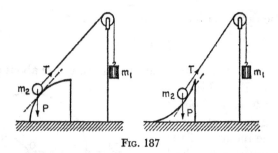

FIG. 187

sider how the forces are altered by a slight displacement of mass m_2 from its position of equilibrium. Then in the first case, if mass m_2 be displaced downwards from its position of equilibrium, the slope of the tangent will be increased and so also will the component of the force of gravity along the rope. It will become greater than the tension T in the rope, and load m_2 will move farther down still. On the other hand, if load m_2 be displaced a small distance upwards, the slope of the tangent will be decreased and so also will the component of the force of gravity along the rope, becoming less than the tension T in the rope; the load will rise farther still. Consequently mass m_2 will not return to its position of equilibrium when it is displaced a short distance from it, but will move farther away, i.e. the equilibrium is unstable.

In the second case, if we displace mass m_2 downwards from its position of equilibrium, the slope of the tangent will decrease, and so the component of the force of gravity along the rope will also decrease; it will become less than the tension in the rope T, and the rope will pull mass m_2 back to its former position. The same reasoning will show us that if the mass m_2 be displaced upwards from its position of equilibrium it will again return to its position of equilibrium, i.e. in this case, equilibrium is stable.

72. Let force F be applied to the left handle of the board. It produces a reaction in the sides of the sideboard at points A and B (Fig. 188). Each of these reactions can be resolved into two components: N_1 and N_2, normal to the sides of the sideboard, and T_1 and T_2, acting along the same sides of the sideboard (the

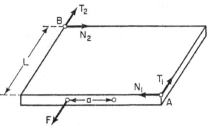

FIG. 188

force of friction). Assuming that the board cannot be pulled out, we must have the following equalities. The force F must equal the sum of the forces of friction, so that there should be no motion in a straight line on the part of the board, i.e. $F = T_1 + T_2$; and the moment of the force F about the centre of the board must equal the sum of the moments of the normal reactions about the same point, for the board not to rotate, i.e.

$$F \frac{a}{2} = (N_1 + N_2) \frac{L}{2}.$$

Further, by definition, we have:

$$\frac{N_1}{T_1} = \frac{N_2}{T_2} = k.$$

Getting rid of force F from both equations, we find that the least value for the coefficient of friction must be L/a. If it has a greater value than this, it is impossible to pull the board out of the sideboard by pulling on one handle only.

IV. Work, Power, Energy. The Law of Conservation of Momentum. The Law of Conservation of Energy

73. To overturn a cube about one edge (e.g. AB) or a cylinder, either must be turned so that their diagonal planes $ABCD$ or $KLMN$ (Fig. 189, a and b) occupy a vertical position. For this work must be done to raise the centre of gravity of the body and the work will be greater the higher the centre of gravity has to be raised (since the weight of cube and cylinder are the same).

If the diagonal plane of the cube or cylinder is to occupy a vertical position, the diagonal AD must be rotated through an angle α about the edge AB and the diagonal plane of the cylinder must be turned through an angle β; the cube's centre of gravity will then rise

$$\Delta h_1 = \frac{h}{2} \left(\frac{1}{\cos \alpha} - 1 \right),$$

and the cylinder's centre of gravity will rise

$$\Delta h_2 = \frac{h}{2} \left(\frac{1}{\cos \beta} - 1 \right)$$

(Fig. 189, c and d). Since the heights and weights of the cube and cylinder are equal and their material is the same, their base-areas are also equal, i.e.

$$h^2 = \pi r^2,$$

where r is the radius of the base of the cylinder. Plainly for the cube $\alpha = 45°$. For the cylinder

$$2r = h \tan \beta$$

or

$$4r^2 = h^2 \tan^2 \beta.$$

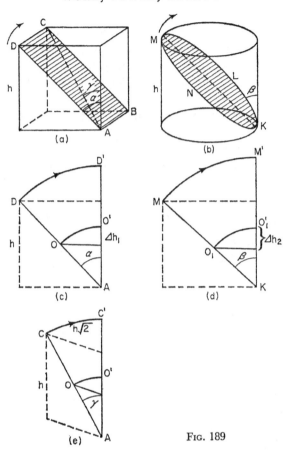

Fig. 189

Substituting for h^2, we get

$$4r^2 = \pi r^2 \tan^2 \beta$$

or

$$\tan^2 \beta = \frac{4}{\pi} > 1,$$

i.e. $\beta > 45°$. Therefore $\cos \beta < \cos \alpha$, and therefore $\Delta h_2 > \Delta h_1$ and it is harder to overturn the cylinder than the cube about one edge.

If the attempt was made to overturn the cube about a corner (instead of about the edge), then it would have to be turned so that the diagonal AC took up a vertical position. For this the cube must be turned through an angle γ, formed by this diagonal and the cube's height (Fig. 189e). Then $\tan^2 \gamma = 2$. Since $\tan^2 B = \frac{4}{\pi} < 2$, $\gamma > \beta$. Therefore $\cos \beta > \cos \gamma$ and it is harder to overturn a cube about one of its corners than to overturn a cylinder.

74. To overturn a crate about edge AB, the crate must be turned so that the diagonal surface shaded in Fig. 190 becomes

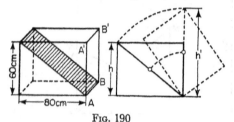

Fig. 190

vertical (see problem 73); from this point the crate will fall as a result of the force of gravity. Thus work must be done in raising the centre of gravity of the crate to the corresponding height. This work equals

$$W = mg\left(\frac{h'}{2} - \frac{h}{2}\right).$$

In overturning the crate about edge AB we have that $h = 0\cdot6$ m, $h' = 1$ m, $W = 0\cdot2$ ton/m. In overturning the crate about edge $A'B'$ we have that $h = 0\cdot8$ m, $h' = 1$ m, $W = 0\cdot1$ ton/m. The total work done is $0\cdot2 + 0\cdot1 = 0\cdot3$ ton/m.

75. If a man moves up an escalator at a constant speed, the average pressure which he exerts on the staircase will remain unaltered and equal to his weight. Therefore the force with which the motor must drive the staircase will also remain the same. However a mounting man will reach the top of the escalator sooner and therefore the distance travelled by the escalator during the period of the man's climb will be less than when the man is stationary. Therefore the work done by the motor of the escalator in raising a moving man will be less than that done in raising a stationary one (the other part of the work is done by the man). The power which has to be exerted by the motor remains the same since the reduced amount of work will be done in a correspondingly reduced period of time.

76. When the pulleys are rotating in a clockwise direction, the belt is in contact with a larger part of each of the pulleys, since the lower part of the belt is pressed on to the pulleys as a result of being stretched, while the upper part of it bends under its own weight (Fig. 191a). Therefore the force of friction will be greater and the belt will begin to slip on the pulleys only when a greater load is applied; therefore more power can be transmitted than when the pulleys are revolving in a counter-clockwise direction (Fig. 191b).

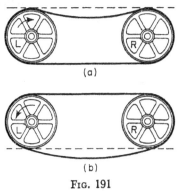

(a)

(b)

Fig. 191

77. For the breech-block to move back a distance a, it is necessary that the work done in overcoming the elastic force of the spring should be $ka^2/2$. This work will be done at the expense of the kinetic energy which the breech-block acquires from the recoil. If the breech-block has initial velocity u, its kinetic energy $Mu^2/2 = ka^2/2$, hence $u = a\sqrt{k/M}$. On the other hand, the absolute momentum Mu of the breech-block on firing must equal the momentum of the bullet, mv (since they are opposite in direction and must give a sum of zero). Therefore

$$v = \frac{Mu}{m} = \frac{a}{m}\sqrt{kM}.$$

78. The modulus of normal elasticity (Young's modulus) is greater for steel than for copper. Therefore if the springs are of equal dimensions and are to be stretched by the same amount, a greater force is necessary for the steel spring than for the copper one. So the first spring requires that more work should be done.

79. (See previous problem.) If the process of stretching is carried out with equal forces for both springs, the steel spring will be stretched less than the copper one. Therefore more work will be done this time on stretching the copper spring.

80. A load m, falling from a height h, acquires a kinetic energy equal to the change in potential energy, i.e. mgh. This kinetic energy must be turned into energy of elastic deformation of the thread, i.e. where Hook's law applies, it must equal $kx^2/2$, where x is the greatest amount of stretch in the thread (at the moment of breaking) and k is the coefficient of elasticity. In the problem it is given that $x = 0 \cdot 01l$ and $kx = Mg$. Substituting these relationships in the equation

$$\frac{kx^2}{2} = mgh,$$

we shall get

$$h = \frac{0 \cdot 01Ml}{2m}.$$

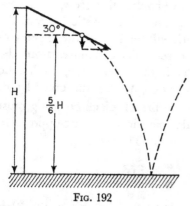

FIG. 192

81. The ball slips from the plane (see Fig. 192) with a velocity of $\sqrt{gH/3}$ at an inclination of 30° to the horizontal. Then the ball describes a parabola and falls on to the horizontal plane with a velocity inclined at some unknown angle to the horizontal. But the height to which the ball will rise after an absolutely elastic impact on the plane depends on only the vertical component of this velocity. The value of this component can be found by calculating the speed with which the ball will fall from a height of $\frac{5}{6} H$ with an initial velocity of $\frac{1}{2} \sqrt{gH/3}$. From the equation

$$\tfrac{5}{6} H = \tfrac{1}{2} \sqrt{\frac{gH}{3}}\, t + \frac{gt^2}{2}$$

we find that the time of fall of the ball

$$t = \frac{\sqrt{21}-1}{2} \sqrt{\frac{H}{3g}}.$$

Therefore its velocity at the end of the fall will be

$$v = v_0 + gt = \frac{\sqrt{21}}{2} \sqrt{\frac{gH}{3}}.$$

Therefore the height to which the ball will rise after its elastic impact on the plane will equal

$$\frac{v^2}{2g} = \frac{7H}{8}.$$

82. A bullet of mass m, travelling with a velocity v, has momentum mv. After the bullet embeds itself in the block, the block plus the bullet will have exactly the same momentum (the impact is completely inelastic). Therefore the velocity v_1, which the block acquires immediately upon the bullet's hitting it, will be determined

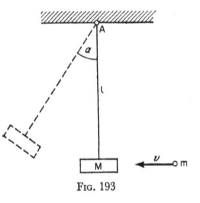

Fig. 193

from the law of conservation of momentum: $mv = (M + m)v_1$. Also the kinetic energy of block and bullet will be

$$\frac{(M + m)v_1^2}{2} = \frac{m}{M + m} \cdot \frac{mv^2}{2}.$$

Then the block will rise, and this kinetic energy will be changed into potential energy. Since the whole mass $(M + m)$ is virtually at a distance of l from the point of suspension A (Fig. 193), its centre of gravity will rise, in consequence of a swing through an angle α on the part of the pendulum, through height $\Delta h = l\,(1 - \cos \alpha)$. At the farthest point of its swing, through an angle α_0, the potential energy must equal the initial kinetic energy, i.e.

$$\frac{m}{M + M} \cdot \frac{mv^2}{2} = (M + m)gl\,(1 - \cos \alpha_0).$$

Hence the angle through which the pendulum swings is given by the relationship

$$\sin^2 \frac{\alpha_0}{2} = \frac{m^2 v^2}{4(M + m)^2 gl}.$$

83. Let the minimum velocity with which the cork must fly out of the test-tube so that the test-tube should describe a full

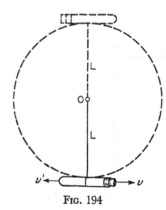

circle about O be v (Fig. 194). Then the cork will acquire momentum of mv, and the test-tube will accordingly acquire momentum of Mv' and the law of conservation of momentum tells us that then $v' = mv/M$. At the initial moment of its movement the test-tube will have a kinetic energy of $Mv'^2/2$. This energy must go towards raising a mass M through a height of $2L$.

From the law of the conservation of energy we shall obtain the equation

$$\frac{Mv'^2}{2} = Mg2L.$$

Hence, substituting for v', we get:

$$v = \frac{2M \sqrt{gL}}{m}.$$

FIG. 194

84. (a) The absolute momentum of the first trolley must equal that of the second trolley since the sum of momentum at the initial moment must equal zero, i.e. $m_1 v_1 = m_2 v_2$; hence

$$\frac{v_1}{v_2} = \frac{m_2}{m_1} = 3.$$

(b) The motion of both trolleys is retarded under the influence of the force of friction. The force of friction may be found from the coefficient of friction and the weight of the trolley

$$f_1 = kP_1 = km_1g, \qquad f_2 = kP_2 = km_2g, \qquad \frac{f_2}{f_1} = \frac{m^2}{m^1} = 3.$$

As a result of the force of friction the velocity of either trolley falls to nought (the trolleys come to a stop). The impulse of the force is equal to the change of momentum.

$$f_1 t_1 = m_1 v_1, \qquad f_2 t_2 = m_2 v_2, \qquad \frac{f_1 t_1}{f_2 t_2} = \frac{m_1 v_1}{m_2 v_2},$$

hence

$$\frac{t_1}{t_2} = \frac{m_1 v_1}{m_2 v_2} \cdot \frac{f_2}{f_1} = 3.$$

(c) The distance travelled by either trolley can be found, when we know the time during which motion takes place and the average speed :

$$s_1 = v_{1_{av}} \cdot t_1 = \frac{v_1}{2} \cdot t_1; \; s_2 = v_{2_{av}} \cdot t_2 = \frac{v_2}{2} \cdot t_2; \; \frac{s_1}{s_2} = \frac{v_1 t_1}{v_2 t_2} = 9.$$

85. Since the first fragment returns, as a result of the explosion, from the highest point of its parabola along its previous trajectory, it follows that it receives, as a result of the explosion, a momentum, mv, which is equal to the momentum which it had before the explosion but with the sign changed ; in other words, as a result of the explosion, a change in momentum of this fragment of the shell has occurred, a change of $-2mv$. By the law of conservation of momentum, the other fragment should also acquire, in the explosion, momentum of the same magnitude, but in the opposite direction, i.e. $+ 2mv$ in the same direction as its previous motion. Therefore the second fragment's momentum after the explosion will equal $3mv$ and it will thus begin its path from the topmost point of its parabola with trebled speed. Consequently it will travel, in a horizontal direction, three times as far as the first fragment, i.e. it will fall to earth twice as far away again as the shell would have fallen, had the explosion in mid-air not take place. Since both fragments had no momentum in a vertical direction immediately after the explosion, they will both fall freely in this direction with no initial velocity and from the same height; therefore they will both hit the ground at the same moment.

Another solution may be put forward, starting from the fact that outside forces play no part during the explosion. Therefore the centre of gravity of the shell has the same velocity both before and after the explosion. Since, in addition to this, both fragments fall freely in the vertical direction after the explosion, the shell's centre of gravity will continue to describe the parabola which the shell would have described, had it not burst. Since the fragments are of equal mass, they will always be symmetrically disposed in relation to the parabola described by the shell's centre of gravity (Fig. 195). Therefore the second fragment will fall at point D, BD being equal to AB.

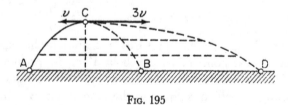

FIG. 195

86. As is well known, if a ball strikes another ball of the same mass, which is stationary, with absolutely elastic, direct impact, the balls exchange velocities, i.e. the stationary one continues to move with the velocity of the first ball, while the latter stops.

But if the direction of velocity of the moving ball does not pass through the centre of the stationary one, then, at the moment of impact, the velocity v of the moving ball can be resolved into two components acting at right angles to one another, one of them passing through the centre of the stationary ball. We can consider the moving ball to have two velocities, v_1 and v_2 (Fig. 196). Along the line of velocity v_1 a direct and absolutely elastic impact takes place, as a result of which the stationary ball begins to move with velocity v_1, while the ball which has been moving previously loses its velocity in this direction, while keeping its velocity v_2, along the line of which it continues to move.

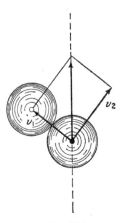

Thus the balls fly apart after impact in directions at right angles to each other.

FIG. 196

87. The impact of balls made of bone may be considered perfectly elastic. If direct and perfectly elastic impact occurs between a moving ball and an identical stationary one, then, as is well known, the stationary ball will begin to move, after impact, with a velocity equal to that which the moving ball had before impact, while the latter stops. In our example a series of successive direct and perfectly elastic impacts of one ball upon another takes place with the same results. But the intermediate balls transmit their velocity to the next ball and themselves stop. The last ball too moves off with the velocity which the first ball had at the moment of impact; but since this last ball does not encounter in its path any other balls, it rises to the same height as that from which the first ball fell (neglecting any losses of energy). Now if not one, but two balls be drawn to the right and released, they will fall independently of one another. On reaching the row of balls 3–8, not one impact, but two will occur, one after the other, i.e. first the second ball will strike the third ball, then the first one will strike the second (Fig. 197). The first of these impacts will lead to the last ball which is free to move (No. 8) breaking contact with No. 7 and beginning to rise. Then the following impact, that of No. 1, will find the seventh ball free to move and it too will start to rise as a result. It is clear that both these balls (Nos. 7 and 8) will swing out as far as balls 2 and 1 were drawn aside to begin with.

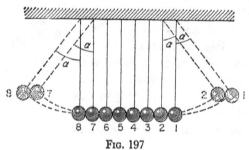

FIG. 197

88. Let us divide both sides of the formula obtained in solving problem 68 by the interval of time t. Then the relationship of the distance from the end of the bar to the interval of time t will be proportional to the velocity. Thus we shall obtain:

$$v_0 = \frac{m_1 v_1 + m_2 v_2 + m_3 v_3}{m_1 + m_2 + m_3}$$

or in general, for any number of spheres

$$v_0 = \frac{\sum_1^n m_i v_i}{\sum_1^n m_i}.$$

89. Use the classic formulas for the velocity of spheres after an elastic impact:

$$v_1' = \frac{(m_1 - m_2)v_1 + 2m_2 v_2}{m_1 + m_2} \text{ and } v_2' = \frac{(m_2 - m_1)v_2 + 2m_1 v_1}{m_1 + m_2}.$$

The velocity of the spheres' centre of gravity after impact, from the formula obtained in the previous problem, will be expressed thus:

$$\frac{m_1 v_1' + m_2 v_2}{m_1 + m_2}.$$

Substitute for v_1' and v_2' in this expression and we shall easily obtain that the velocity of the sphere's centre of gravity after an elastic impact will equal

$$\frac{m_1 v_1 + m_2 v_2}{m_1 + m_2},$$

which is the same as the velocity of the spheres' centre of gravity before impact.

If the impact is inelastic, both spheres move together, i.e. with the same velocity. Therefore their centre of gravity will move with this velocity, which is also expressed by the proportion

$$\frac{m_1 v_1 + m_2 v_2}{m_1 + m_2}.$$

Therefore the same rule holds good for an inelastic impact also. Thus the momentum of the centre of gravity of a system of spheres is the same before and after impact. In particular, if the centre of gravity of the system is at rest before impact, it will also be stationary after impact (whether the impact is elastic or inelastic) What obtains for spheres, obtains also for any system of particles of matter. The law deduced by us here should be considered an alternative formulation of the law of conservation of momentum, i.e. if outside forces do not affect it, the centre of gravity of any system of particles of matter conserves its momentum and therefore also its velocity. This fact sometimes allows some mechanics problems to be solved more simply. We shall demonstrate this later.

90. Since there are no external forces which could impart momentum to the monkeys, it follows that the monkeys can impart to each other (through the rope) only equal momentum.

Therefore, however quickly the monkeys move their paws up the rope, they will rise, relative to the ground, at equal rates, since they are of equal mass. And the rope will move through the pulley, to the side of the monkey which climbs the faster, with a velocity which will make the rates at which the monkeys climb, relative to the ground, equal. Therefore both monkeys will reach the pulley at the same time.

The problem can be simplified so that the solution should be quite clear. Let us imagine that the monkeys are on an absolutely smooth horizontal surface and hold opposite ends of a rope. Since there are no external forces acting, the centre of gravity of the two monkeys must remain stationary and therefore they can only move equal distances towards this centre of gravity, however fast either of them moves its paws along the rope. Therefore both monkeys will reach the point lying at the centre of the original distance between them simultaneously.

This same reasoning allows us to answer easily enough, if the problem be posed with monkeys of different mass: the lighter will reach the top first.

91. Suppose that the man moves from bow to stern uniformly and in time t. Since there are no external forces, the momentum of the system boat-plus-man cannot change, i.e. throughout the man's movement, the boat must move in the opposite direction with a velocity such that the total momentum of the system equals zero. Let the boat, in the same time t, move distance x in the opposite direction (Fig. 198). Then the velocity of the man relative to the ground during this period was $(L-x)/t$, and the velocity of the boat was x/t. The law of conservation of momentum gives us that

$$\frac{m(L-x)}{t} - \frac{Mx}{t} = 0,$$

hence

$$x = \frac{mL}{M+m} = \frac{60 \cdot 3}{180} = 1 \text{ m}.$$

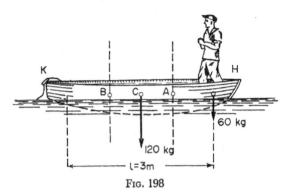

Fig. 198

The same result can be obtained by starting from a consequence of the law of conservation of momentum, that if there are no external forces acting on a system, the system's centre of gravity must remain stationary. When the man is standing in the bow H of the boat, the centre of gravity of boat-plus-man is on a vertical line passing through A, such that $CA = 0.5$ m. When the man has moved to the stern K, the centre of gravity of the system lies on the vertical line passing through B, where $BC = 0.5$ m. Since no external forces were acting on the system during the man's progress from bow to stern, the centre of gravity of the system can not have moved. For this to be so, the boat must move in such a way that point B should coincide with the previous position of point A, i.e. the boat must move a distance BA to the right, a distance of 1 m.

92. Clearly when both guns fire simultaneously, the truck will remain stationary and under these conditions (firing from a stationary platform), the shells hit their target. If one of the guns fires sooner than the other, then, in accordance with the law of conservation of momentum, the truck will begin to move in the opposite direction. Since all the conditions governing the motion of the shell remain the same in the barrel of the gun, while the barrel itself begins to move, together with the truck, in the opposite direction (and this motion begins in conjunction with the shell, i.e. while it is still in the barrel of the gun), the shell will leave the barrel with a velocity relative to the earth, which is reduced by comparison with its velocity when fired from a stationary platform. Therefore the shell will fall short of its target. The platform's motion will continue right up to the firing of the second gun. Therefore, when it is fired, the second gun's barrel will be moving, together with the truck, in the same direction as its shot; thus the shell will have a slightly greater velocity when leaving the gun than if it had been fired from a stationary gun. Therefore it will overshoot its target.

Since the velocity of the second shell when fired is greater than that of the first, it follows from the law of conservation of momentum that after the second shot is fired the truck must begin to move in a direction opposite to the direction of the shot. For simplification, we do not take into account the circumstance that after firing the gun's barrel will move relative to the gun-carriage. But it is clear that this will not, in principle, change the part played by the truck's movement.

93. Springs *I* and *II*, having equal elasticity, act with different forces of $2mg$ and mg at the initial moment, and yet they have the same length. Thus, when they are not under stress, they must have different lengths. In a free fall, both springs must cease to be stretched, i.e. they will contract to their normal size (their deformation will disappear) and since this normal length is not the same for the two springs, the distances between the centres of the first and second ball and the second and third ball will no longer be the same. Thus the centre of the second ball will cease to be the centre of gravity of the system of three balls after the beginning of the fall.

94. A watch's hair-spring is attached by one end to the body of the watch and by the other to the balance-wheel (Fig. 199). The period of oscillation of the balance-wheel depends on the elasticity and length of the spring. The shorter the spring, the smaller the period of oscillation. When the watch lies on the stand we must assume that the body of the watch is firmly attached to it. If the stand can rotate freely about its vertical axis, a turn of the balance-wheel in one direction will cause the stand and the body of the watch to rotate in the opposite direction. As a result of this the spring will be tightened from both ends in opposite directions. Therefore the stationary point on the spring will be somewhere inside the spring, between the point of attachment to the watch-body and the balance-wheel. In other words, the length of the spring is, as it were, reduced and this leads to a reduction in the period of oscillation. So the watch will start to gain.

Fig. 199

95. A graphical explanation is most easily understood (Fig. 200). Since all three balls have the same initial velocity, the distance-time graph for each one (distance here being height) will look the same, supposing that no collisions occur. To bring into account the result of the collisions, the following points should be considered: (1) the balls collide when the graphs of their motion intersect, (2) on collision, the balls exchange velocities (elastic impact), i.e. after collison, each ball continues

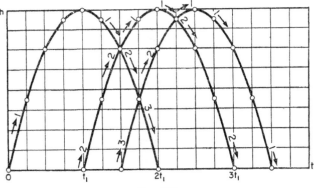

Fig. 200

the motion of the other (i.e. along the other's graph). (In Fig. 200 the figures indicate which graph refers to the motion of which ball.) We shall reckon the lapse of time from the moment of throwing up the first ball, and we shall call the time needed for it to reach its highest point t_1. Then the third ball will fall after $2t_1$, the second after $3t_1$ and the first after $3\frac{1}{2}t_1$.

96. The principle of relativity demands that the same physical laws should obtain in the two systems under consideration, and in particular the law of conservation of energy, according to which the change of energy of a body must equal the work done by external forces. Therefore in system *I* the following relationship must obtain

$$\frac{m}{2}\,(v_2{}^2 - v_1{}^2) = Fs, \qquad (1)$$

where s is the distance travelled by a body in system *I* in the time during which its velocity rises from v_1 to v_2.

Correspondingly in system *II*

$$\frac{m}{2}\,(v_2{}^2 - v_1{}^2) - mv(v_2 - v_1) = Fs_1, \qquad (2)$$

where s_1 is the distance travelled by a body in system *II* in the same time. But since the velocity of a body in systems *I* and *II* is not the same, neither are s and s_1 the same. In fact, since a body moves, under the impulse of a force F, with acceleration F/m, in system *I* we have:

$$s = v_1 t + \frac{F}{m} \cdot \frac{t^2}{2},$$

and in system *II* correspondingly

$$s_1 = (v_1 - v)t + \frac{F}{m} \cdot \frac{t^2}{2}.$$

Therefore $s - s_1 = vt$. But since $F/m = a = (v_2 - v_1)/t$,

$$t = \frac{v_2 - v_1}{F} \cdot m$$

and

$$s-s_1 = \frac{v(v_2 - v_1)}{F} \cdot m,$$

and therefore $F(s-s_1) = mv(v_2-v_1)$.

Thus the work of outside forces in system I is as much greater than the work of outside forces in system II as the change in kinetic energy in system I is greater than the change in kinetic energy in system II. And since the change in energy in system I equals the work done by external forces, this also holds for system II. Therefore Newton's principle of relativity is not broken.

V. The Dynamics of Motion in a Circle

97. The force of friction in both cases is the same, but the moment of the force of friction relative to the axis is less in the case of the shaped shaft, and the loss due to friction in rotation is determined precisely by this moment of the force of friction.

98. Let the unknown force which one brake exerts on the wheel be P. The force of friction will equal kP. The moment of the forces of friction acting on the two brakes will equal $kP.2R$. This moment of the forces of friction must balance the moment of the rotating couple M. Therefore $2kPR = M$. Hence $P = M/2kR = 400$ kg.

99. If the pulley were weightless, we should consider the tension in the rope on both sides of the pulley to be the same (see problem 42). But since the pulley has a moment of inertia, we must take into account that the tension in the rope on the two sides of the pulley will be different and this difference of tension will create the moment which rotates the pulley. Let the tension in the rope on the left of the pulley be T_1 and the tension in that on the right be T_2. Then, applying Newton's second law to the motion of masses m_1 and m_2, we shall obtain the equations

$$m_1 a = m_1 g - T_1, \tag{1}$$
$$-m_2 a = m_2 g - T_2. \tag{2}$$

Here we have considered as positive the direction of acceleration of load m_1 and so acceleration a in the second equation is given the minus sign.

If the angular velocity of the pulley be denoted by ϵ, so that $\epsilon = a/R$, where R is the radius of the pulley, then applying Newton's second law to its rotation we shall have

$$I\epsilon = (T_1 - T_2)R. \tag{3}$$

Solving equations (1), (2) and (3) simultaneously, we shall have:

$$T_1 = m_1 g \, \frac{2m_2 + IR^2}{m_1 + m_2 + IR^2},$$

$$T_2 = m_2 g \, \frac{2m_1 + IR^2}{m_1 + m_2 + IR^2}.$$

When $I = 0$ we obtain from these equations that $T_1 = T_2$, which coincides with the results of problem 42.

100. At the moment that the front wheels are turned, the car possesses velocity in a straight line v, which may be resolved in two directions x and y at right angles to one another (Fig. 201). The

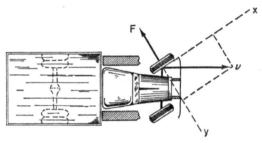

FIG. 201

wheels should simultaneously roll along x and slip along y. But if slipping ought to take place, then the force of friction always comes into play, acting in the opposite direction to that in which slipping should take place. This external force F is what causes the whole car to turn the way required.

101. Since the centripetal force at the upper part of the loop is that which pulls the body towards the earth, the centrifugal force is that with which the earth is pulled towards the body, i.e. it acts on the earth.

102. The plumb-line is so set up that the resultant of its weight mg and the tension in the thread T produces a centripetal force $F = m\omega^2 R$ (Fig. 202). Clearly $R = r + l \sin \alpha$. Therefore,

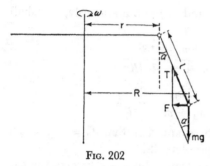

FIG. 202

$$\omega^2 = \frac{g \tan \alpha}{r + l \sin \alpha},$$

$$\omega = \sqrt{\frac{g \tan \alpha}{r + l \sin \alpha}}.$$

103. The force of gravity P acting on a coin rolling in a vertical position and the reaction R of the plane balance one another and therefore they cannot bend the path which the coin follows (Fig. 203a). When the coin is tilted (Fig. 203b), the normal reaction R_1 of the plane equals the weight of the coin as before; but, in addition to this, the

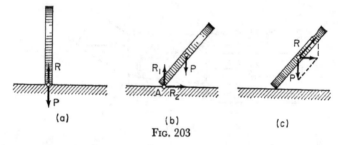

(a)   (b)   (c)

FIG. 203

force of friction R_2 comes into play, acting along the plane. Its appearance in the picture can be explained by the fact that in its inclined position the coin will be subject to slipping at point A, to the left. But at the same time the force of friction always opposes the occurrence of this slipping. Thus R, the resultant of the forces of reaction R_1 and R_2 of the plane, is inclined to the plane (Fig. 203c). The resultant of the force of gravity and the force of the reaction R acts horizontally in the same direction as the tilt and causes the trajectory of the centre of gravity of the coin to curve.

104. Two forces are acting on the skater as he leans towards the centre of the circle: his own weight and the reaction of the ice-track, which has a horizontal component and is therefore inclined at an angle. The horizontal component of the ice-track's reaction (Fig. 204) comes into play as a result of the fact that the skater, gliding along his path, does not glide in a perpendicular position. This horizontal component is analogous to the force of friction which acts on the coin in problem 103.

The resultant of the weight of the skater P and the inclined

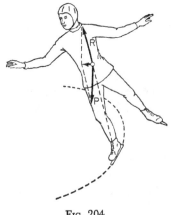

Fig. 204

reaction of the ice-track R is a force acting horizontally and imparting to the skater the centripetal acceleration necessary to motion in a circle.

105. For the pilot to describe the given loop, he must be subject to a centripetal force at all points on the loop of

$$\frac{mv^2}{R} = 108 \cdot 5 \text{ kg.}$$

At the lowest point in the loop, this force comes from a part of the seat's reaction, and at the highest point from the pilot's weight plus the reaction of the seat. Therefore the centrifugal force at the lowest point of the loop acts entirely on the seat, and at the highest point partly on the seat and partly on the earth. At the lowest point of the loop, both the pilot's weight and the centrifugal force are acting on the seat, so the pilot is pressed to his seat with a force of 178·5 kg; and at the highest point of the loop only part of the centrifugal force is acting on the seat, equal to the centrifugal force minus the wight of the pilot, so the pilot at the highest point of the loop is pressed to his seat with a force of 38·5 kg.

106. If the sphere has been raised a height h, its potential energy has increased by mgh. As the pendulum swings, this energy will be transformed into kinetic energy and back again;

and when the sphere rises to its highest point again and comes to rest it will have the same potential energy. Thus the ball must rise to the same height as before, since the fact that the thread meets the bar in no way alters the kinetic energy of the sphere. Therefore, wherever the bar be placed at right angles to the plane of the sketch, provided the distance AB be less than $l-h/2$, the sphere will rise to the same height h. But if AB be greater than $l-h/2$, then this will not be possible. But in this case when the sphere reaches its highest point it will have some velocity still (since not all its kinetic energy will have been changed into potential energy). It will continue to move in the same direction and the thread will wind itself round the bar.

107. The ball first of all describes a quadrant of a circle of radius equal to the length l of the thread. Then the thread touches the nail O, which was driven into the wall, and the ball

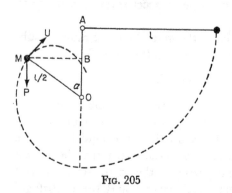

FIG. 205

describes an arc of a circle of radius twice as small as the previous one. Finally, when the weight of the ball gives it the centripetal force necessary to motion in a circle, the tension in the thread will be reduced to zero. Let this happen at point M (Fig. 205). We can find this point's position in the following manner. The component of the ball's weight acting along the line of the radius equals $P \cos \alpha$, where α is the angle made by the thread at that moment. Further, at point M, v^2 equals $2gh$, where $H = AB = AO - BO = l/2 - l\cos\alpha/2$. Therefore the centripetal force at point M equals

$$\frac{mv^2}{R} = \frac{2m\left(\dfrac{l}{2} - \dfrac{l}{2}\cos\alpha\right)g}{l/2} = 2P(1 - \cos\alpha).$$

Thus at point M we have the equation $P \cos \alpha = 2P(1-\cos \alpha)$, from which $\cos \alpha = 2/3$. The ball continues its path just as a body thrown at an angle of α to the horizontal with an initial velocity $v = \sqrt{gl/3}$. In this case, the highest point of the parabola lies above the point of throwing at a distance of

$$\frac{(v \sin \alpha)^2}{2g} = \frac{5l}{54}.$$

The vertical line which passes through the point of suspension lies at a distance from M of

$$MB = \frac{l \sin \alpha}{2} = \frac{\sqrt{5}}{6} l.$$

To travel this distance horizontally the ball will require time

$$t = \frac{MB}{v \cos \alpha} = \frac{\sqrt{15}}{4} \cdot \sqrt{\frac{l}{g}}.$$

During this time the ball will travel, in height, a distance of

$$v \sin \alpha \, t - \frac{gt^2}{2} = \frac{5l}{96},$$

i.e. it will cut the vertical line AO at a point lying $5l/96$ below point B.

108. The ball will move more slowly if the force of friction is directed against the line of motion and faster if it is acting along the direction of motion. The direction of the force of friction is opposite to the direction of the velocity with which slipping takes place at K—the point of contact of the ball with the

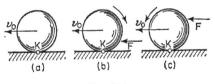

(a)   (b)   (c)

FIG. 206

surface (Fig. 206a). The velocity of slipping at this point of contact depends on the relationship between two velocities—the velocity of the motion in a straight line of the ball, v_0 and the velocity of rotation about the centre of gravity.

If the ball be struck low (the ball is rotating clockwise,

Fig. 206b) the velocity at point K determined by the rotation, is in a forward direction and slipping occurs, also in a forward direction. Therefore the force of friction comes into play, acting in a backward direction, and this slows down the ball's motion.

If the ball be struck high (Fig. 206c), the ball rotates in a counterclockwise direction. If the ball be struck high enough, when its velocity of rotation is sufficiently great (so that the linear velocity caused by this rotation at point K is greater than the velocity of the ball's motion in a straight line) slipping occurs in a backward direction. The force of friction comes into play, directed along the line of motion, and the ball moves faster.

109. If the bullet be considered from outside, not from the earth but, e.g. from the moon, its path will appear to be a vertical fall to earth, since the velocity of the bullet relative to the earth and the velocity of the earth relative to the outside observer are equal and directly opposite.

If the bullet be observed from the earth its path will appear to be a parabola, as normal, ending in the bullet's fall to earth.

110. Rolling without slipping can be considered as rotation about a momentary axis O (the part of the cylinder touching the plane at any given moment) together with the linear motion of this axis. Then the turning effect is the moment of the force of gravity P relative to the axis O (Fig. 207). The cylinders' masses, and therefore their weights, are equal and so the turning effect is the same for both cylinders. But their moments of inertia are different.

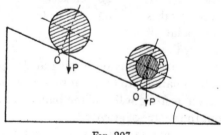

FIG. 207

In the second (compound) cylinder only the moment of inertia of the outside cylinder plays a part, since the inside cylinder, in the absence of the force of friction, does not rotate. Obviously the moment of inertia of the first (solid) cylinder is greater than the moment of inertia of the outside, hollow cylinder in the second case. Given the same turning effect the angular velocity is in inverse proportion to the cylinder's moment of inertia. Therefore the angular velocity of the solid cylinder is less, i.e. it rolls down more slowly.

The position of the inside cylinder may be determined from the following considerations. If the inside cylinder were to slip down an inclined plane without friction, its acceleration from the effect of the force of gravity would be greater than the acceleration of the centre of a cylinder which rolls down. This can be seen from the fact that the potential energy which the cylinder has in the gravitational field is turned, in slipping, into only kinetic energy of linear motion, but in rolling it is turned into energy of both linear and rotatory motion. Consequently the velocity, and therefore the acceleration, of a cylinder's linear motion during slipping must be greater than during rolling. Since in

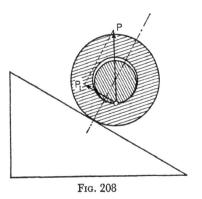

FIG. 208

the case of the compound cylinder both the outside and the inside cylinder move downwards with identical acceleration, this means that the outside cylinder retards the motion of the inside one. For this, force P, which the outside cylinder exerts on the inside one must have a component P_1 which acts along the line of slope of the plane in an upward direction, i.e. force P must be inclined somewhat backwards of the perpendicular to the inclined plane. Since there is no friction, force P must be normal to the surface of the cylinders; for this to be so when it is inclined backwards, the cylinders must be in contact along a line which lies at least in front of the radius drawn to the point of contact between the outside cylinder and the inclined plane (Fig. 208).

VI. The Universal Theory of Gravitation

111. If the sphere were solid it would attract the small ball with a force

$$F = \gamma \frac{Mm}{d^2},$$

where γ is the gravitational constant. We may consider that the attractive force F exerted by the solid sphere is composed of two forces: the attractive force exerted by the smaller sphere of radius $R/2$ and the attractive force exerted by the remainder of the larger sphere. The problem requires us to find the second of these forces. The mass of the sphere which would fill the hollow is

$$M' = \frac{4}{3}\,\pi\rho\left(\frac{R}{2}\right)^3 = \frac{M}{8},$$

and its centre lies at a distance $d - R/2$ from m. The unknown force which equals the difference between the forces of the whole sphere and of the smaller sphere hollowed out, will be expressed thus:

$$F' = \gamma \frac{Mm}{d^2} - \gamma \frac{\frac{M}{8}m}{\left(d - \frac{R}{2}\right)^2} = \gamma\, Mm\left[\frac{7d^2 - 8dR + 2R^2}{8d^2\left(d - \frac{R}{2}\right)^2}\right].$$

Many who sat for the physics paper in the "Olympic" examination of Moscow State University for 1946, in which this problem was set, gave another and wrong solution. Since their mistake is instructive, we will give this solution and explain the fault in it.

The position of the new centre of gravity of the sphere after hollowing was calculated; the distance from it to the centre of the whole sphere can be found from the equation

$$Mgx = \frac{Mg}{8}\left(\frac{R}{2} + x\right),$$

from which $x = R/14$. Then the attractive force exerted by the hollowed sphere on the ball of mass m was found (the mass of the hollowed sphere is $\frac{7}{8}$ of the mass of the original sphere) as though we had two particle-masses at a distance of $d + R/14$ from one another, i.e. from the formula

$$F = \gamma\, \frac{\dfrac{8}{7}Mm}{\left(d + \dfrac{R}{14}\right)^2}.$$

It is not difficult to see that this result is different from the one which was obtained in our solution. They are the same only for values of $R \ll d$. The mistake lies in the incorrect assumption that a hollowed sphere attracts a mass m in the same way as a particle of the same mass would do if it lay at the centre of gravity of the hollowed sphere.

The centre of gravity is the point of application of the resultant of all parallel forces acting on distinct elements in a body, and the size of each of these forces is proportional to the mass of a given element of the body. But the forces with which the various elements of a body act on a mass m are, first of all, not parallel, since they are all directed towards point m, and, second, although they are proportional to the masses of the elements of the body, yet they are in general different for elements of equal mass, since they depend on the distance of the given element from the mass m. Therefore it is generally speaking impossible to substitute, for the force of gravitation of a given body, the force of gravitation of a particle of the same mass lying at the centre of gravity of the given body. Only in special cases, when the dimensions of the body are small by comparison with the distance between them (i.e. when the bodies can be treated as particles of matter), or when the attracting body is of a particularly symmetrical shape, e.g. a homogeneous sphere, can the force of gravitation of such a body be calculated as if all its mass were concentrated at its centre of gravity. It was this last point that allowed us to calculate as we did the force of attraction of the solid sphere and of the sphere that was removed.

112. The phenomenon of ebb and flow arises as a consequence of the fact that a given body in the heavens (the moon, the sun) imparts different accelerations to the earth as a whole and to the water which is on the earth's surface. A heavenly body imparts to the earth as a whole the same acceleration as it would to a body

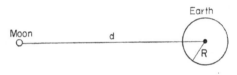

Fig. 209

located at the centre of the earth. But according to the universal theory of gravitation the acceleration imparted by a body of mass M to another body distance r away is $a = \gamma M/r^2$, where γ is the gravitational constant. Consequently the difference between the acceleration of the water on the earth's surface and the acceleration of the earth as a whole will be expressed thus:

$$\frac{\gamma M}{(d-R)^2} - \frac{\gamma M}{d^2} = \frac{\gamma M(2dR - R^2)}{d^2(d-R)^2}.$$

where d is the distance from the heavenly body to the centre of the earth, and R is the radius of the earth (Fig. 209). Since the distance R is very small by comparison with d, we shall have an approximation for this difference of

$$2R\frac{\gamma M}{d^3}.$$

158   SOLUTIONS

The value of this difference for the moon and for the sun is what
determines the tidal effect each of them has. Since

    d for the moon is approx. 60 times the radius of the earth,
    d for the sun is approx. 2500 times the radius of the earth,

d^3 will be, for the sun, approx. 75×10^6 times as great as it will
be for the moon, while the sun's mass is only approx. 27×10^6
times greater than the moon's. Therefore the effect of the moon
on tides is almost three times as great as that of the sun.

113. The force of attraction exerted by the sun acts not only
on a load in a pair of scales, but also on the earth, and therefore
it imparts identical acceleration to both scales and load (neglecting tidal effects, see preceding problem). Therefore the force
of attraction of the sun does not alter the strain in the springs, i.e.
it does not affect the reading of the scales. Just as the earth's
attractive force does not stretch the spring if the load, together
with scale and spring, falls freely to earth, so the sun's attractive
force does not stretch the spring, since the scale "falls", together
with the earth towards the sun. This "fall" expresses itself in the
fact that the earth, in its orbit round the sun, has a centripetal
acceleration, which is in fact equal to the acceleration of free fall,
which it or any other body would have if placed on the earth's
orbit without initial velocity. Consequently the weight of bodies
on the earth is the same day and night.

114. (1) According to the universal theory of gravitation, the
acceleration acquired by a body is in inverse proportion to the
distance squared. Therefore a body which has been raised to a
height of 500 km and which therefore is at a distance of 7000 km
from the centre of the earth, will have a gravitational acceleration of

$$\frac{980 \times 6500^2}{7000^2} = 845 \text{ cm/sec}^2.$$

(2) For the body to be able to describe a circle about the earth,
the acceleration imparted to it by the earth must be equal to the
required centripetal acceleration. Therefore

$$845 = \frac{v^2}{7000 \times 10^5}.$$

Hence we get that $v = 7\cdot7$ km/sec.

(3) At the velocity we have found, one revolution about the earth will be completed in a time equal to

$$\frac{2\pi \times 7000}{7\cdot7 \times 3600} \text{ hr,}$$

or approximately 1 hr. 30 min.

VII. Oscillation, Waves, Sound

115. As long as Hooke's law obtains, the displacement of mass m a distance x from the position of equilibrium gives rise to a force

$$f = -(k_1 + k_2)x,$$

which does not depend on how much the springs were extended when mass m was in a state of equilibrium. Since the dependence of the force acting on m upon the amount of displacement from the position of equilibrium determines the period of oscillation, and since this dependence is not altered by any change in the extension of the springs required by equilibrium, the period of oscillation will not depend on this extension either; consequently the moving of the points of attachment from A_1 and A_2 to B_1 and B_2 (Fig. 60) will only lead to a change in the location of the central position (about which oscillations take place) but will not affect the period.

116. When mass m reaches the scale-pan, it will have a kinetic energy of

$$\frac{mv^2}{2} = mgh. \tag{1}$$

After impact the load and the pan together will have the same momentum as the load had before impact, and (since we are ignoring the mass of the scale-pan) the same velocity and the same kinetic energy, viz., mgh. After impact the scale-pan plus load will move downwards and extend the spring. The extension of the spring will take place at the expense of the initial kinetic energy and the work done by the force of gravity. If the displacement downwards of the pan be considered positive, then the work done by the force of gravity is mgx, where x is the pan's displacement from its initial position. Therefore the greatest extension x_0

of the spring is determined by the fact that all the kinetic
energy and the work done by the force of gravity go into the
elastic deformation of the spring, i.e.

$$\frac{kx_0{}^2}{2} = mgh + mgx_0 \qquad (2)$$

or

$$x_0{}^2 - \frac{2mg}{k}\,x_0 - \frac{2mgh}{k} = 0.$$

hence

$$x_0 = \frac{mg}{k} + \sqrt{\frac{m^2g^2}{k^2} + \frac{2mgh}{k}}.$$

Here the positive root of this equation corresponds with the
lowest displacement downwards (since we agreed to consider
downward displacement as positive). Since this greatest displacement is greater than mg/k (which corresponds with the scale-
pan's position of equilibrium when the load is in it), it follows
that the pan, having reached its lowest position will begin to rise,
will pass through its original position and will rise farther,
compressing the spring. When it comes to rest at its highest
position the potential energy of the compressed spring will
again equal the sum of the initial kinetic energy and the work
done by the force of gravity, i.e. the greatest displacement upwards will also be determined by equation (2), but the
second, negative root will correspond with this upward displacement. And so the scale-pan will make oscillations between the
two extreme positions

$$x_{01} = \frac{mg}{k} + \sqrt{\frac{m^2g^2}{k^2} + \frac{2mgh}{k}}.$$

and

$$x_{02} = \frac{mg}{k} - \sqrt{\frac{m^2g^2}{k^2} + \frac{2mgh}{k}}.$$

And to the position of equilibrium of the scale-pan will correspond a displacement of

$$x_0 = \frac{mg}{k}.$$

Consequently the greatest displacements in either direction from the equilibrium position will be identical, and equal

$$\sqrt{\frac{m^2 g^2}{k^2} + \frac{2mgh}{k}}. \qquad (3)$$

This is the amplitude of the scale-pan's oscillations.

If we are not allowed to neglect the mass of the scale-pan M, then the velocity with which the scale-pan will begin to move downwards will not equal the velocity with which the load reaches the scale-pan and which is determinable from equation (1). To find the velocity V with which the scale-pan begins to move downwards under the influence of the load's fall, we must make use of the law of conservation of momentum. In this case we shall have

$$mv = (M + m)V \text{ and } V = \frac{m}{M + m} v.$$

Substituting for v from equation (1), we shall get:

$$V = \frac{m}{M + m} \sqrt{2gh}.$$

Further it should be remembered that the moving mass will now be $M + m$. Besides this, at the initial moment the spring is extended a distance a as a result of the weight of the scale-pan, so that

$$ka = Mg \qquad (4)$$

Therefore the law of conservation of energy gives

$$\frac{kx^2}{2} - \frac{ka^2}{2} = \tfrac{1}{2}(M + m) \left[\frac{m\sqrt{2gh}}{M + m}\right]^2 + (M + m)g(x - a).$$

Substituting for a in this equation from equation (4) and taking all the terms over to the left-hand side, we shall obtain:

$$x^2 - \frac{2(M+m)g}{k} x - \frac{2m^2gh}{(M+m)k} + \frac{M(M+2m)g^2}{k^2} = 0.$$

Hence

$$x = \frac{M+m}{k} g \pm \sqrt{\frac{m^2g^2}{k^2} + \frac{2m^2gh}{(M+m)k}}.$$

The new equilibrium position will be

$$x_0 = \frac{M+m}{k} g.$$

Reasoning analogous to the previous case leads to this value for the amplitude

$$\sqrt{\frac{m^2g^2}{k^2} + \frac{2m^2gh}{(M+m)k}}. \qquad (5)$$

As we should expect, expression (5) becomes (3), if we put $M = 0$.

117. As the liquid flows out of the vessel, first of all the centre of gravity of the liquid and so too the centre of gravity of the pendulum will move downwards and the distance from the centre of gravity to the point of suspension will increase. Therefore at first, as the water flows out, the period of oscillation of the pendulum will gradually grow.

However the lowering of the vessel's centre of gravity does not take place consistently. When there is little water only left in the vessel, the centre of gravity of the vessel may begin to rise as this small quantity flows out and the period of oscillation decrease. This can be seen from the fact that when all the water has flowed out of the vessel, the pendulum's centre of gravity will be higher than when the level of water in the vessel is somewhat below the centre of gravity of the vessel itself. A completely consistent change of period of oscillation will only occur if the centre of gravity of the vessel itself is located in the bottom of the vessel.

118. The period of oscillation will be less in the second case. For in the first case each pendulum oscillates only under the influence of its own weight, since the spring neither extends nor contracts (Fig. 210). In the second case the force of the spring's elasticity is added to the component of the force of gravity and acts always towards the equilibrium position. Therefore the acceleration of the pendulums at any moment in the second case will be greater than in the first. Since the period of oscillation is smaller, the greater the acceleration, it follows that the period of oscillation is less in the second case than in the first.

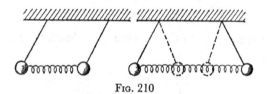

FIG. 210

119. As the pendulums were first drawn to the same side through the same angle, they will make their oscillations also in such a way that at every moment their inclination will be the same (i.e. the oscillations will occur with identical phases). Then the spring will not be extended and, if it is weightless, will have no effect on the pendulums' period of oscillation : each pendulum oscillates with a period which belongs to it separately. The mid-point of the spring also makes oscillations in this case. But if the spring be fixed at its mid-point, the spring will begin to affect the pendulum's period of oscillation, since it will be extended and compressed during oscillation. The pendulum's oscillations will take place not only under the influence of its own weight, but also under the influence of the spring's elasticity, which always acts with a force directed towards the position of equilibrium, and therefore increases the pendulum's acceleration by comparison with that which it would have under the influence of only the force of gravity. This leads to a reduction of the period of oscillation.

120. An object lying upon the stand will not be separated from it at the stand's highest position, if the stand's acceleration downwards in its harmonic oscillation is not greater than the acceleration due to the force of gravity. The stand will have the greatest downward acceleration when it is at its highest point (Fig. 64). If the amplitude of the oscillations is A and the period T, the amplitude of the acceleration will equal $A(2\pi/T)^2$. We shall obtain the least value for the period of oscillation T, by making the amplitude of the stand's acceleration equal to the acceleration of the force of gravity, i.e.

$$A\left(\frac{2\pi}{T}\right)^2 = g,$$

hence

$$T = 2\pi\sqrt{\frac{A}{g}} = 2\pi\sqrt{\frac{0\cdot5}{9\cdot8}} = 1.4 \text{ sec approx.}$$

121. A body resting on the stand will not slip across it until the force of friction acting on the body on the part of the stand is sufficiently great to impart to the body the same acceleration as the stand has, i.e. till it is greater than the product of the mass of this body and the maximum value for the stand's acceleration. If the oscillation has an amplitude of A and period T, the amplitude of acceleration will equal $A(2\pi/T)^2$. Consequently the body will remain at rest so long as the force of friction $F \geqslant mA(2\pi/T)^2$.

On the other hand if mg is the weight of the body and k is the coefficient of friction, then this same force of friction $F = kmg$. Therefore the body will begin to slip when

$$k = \frac{A\left(\frac{2\pi}{T}\right)^2}{g}.$$

Substituting the numerical quantities given, we shall obtain:

$$k = \frac{0\cdot6\left(\frac{2\pi}{5}\right)^2}{9\cdot8} = 0\cdot1 \text{ approx.}$$

122. To assess the effect on the character of a pendulum's motion of accelerating the point of suspension, we can make use of the analogy of the behaviour of a spring-balance (dynamo-meter). As you know, if the point of suspension of a spring-balance be accelerated upwards with acceleration a, the balance registers force $F = m(g + a)$, where m is the mass of the attached load; if the acceleration is downwards, $F = m(g-a)$. Con-

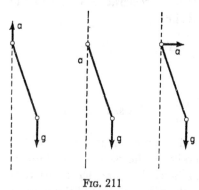

FIG. 211

sequently, the acceleration of the point of suspension is equivalent to the appearance of an additional force ma, acting in the direction opposite to that of the acceleration (thus, when the acceleration is upward, the force acts downwards and is added to mg, if the acceleration acts downwards, the force is subtracted from mg).

The acceleration of the point of suspension will have just the same effect on a pendulum's oscillations (Fig. 211).

(1) The bob's acceleration will be greater at every point than when the point of suspension is stationary and the period will be reduced.

(2) The bob's acceleration will be less at every point than when the point of suspension is stationary, and the period will be increased.

(3) The force of gravity and the additional force to which the point of suspension's motion is equivalent act in directions at right angles to each other, and their resultant will be inclined at an angle α to the vertical, on the side opposite to the acceleration a; angle α may be found from the relationship $\alpha = g/a$ (Fig. 211). Therefore, if the pendulum is not oscillating it will hang inclined at an angle α to the vertical away from the direction of acceleration a. If the pendulum be drawn out of this equilibrium position it will oscillate about it with a period corresponding with the acceleration $\sqrt{a^2 + g^2}$, i.e. greater than g. Con-

sequently the period of oscillation of the pendulum will be reduced.

123. If the board and pendulum begin to fall freely, the force of gravity acting on the pendulum imparts to it the acceleration of free fall and no longer acts as a force tending to return the pendulum to the position of equilibrium as it did when the pendulum was attached to the stationary board. Since the board is massive, the motion of the pendulum does not affect its motion, so the board has the acceleration of free fall. Since there are no other external forces, the pendulum will preserve the motion relative to the board which it had at the beginning of the fall, as though there were no force of gravity. Consequently: (1) the bob will remain stationary relative to the board in its rest-position (since its velocity in this position equals zero), (2) the bob will continue to oscillate uniformly about the point of suspension with the same velocity as it had at the initial moment of fall. This same result can be reached with the help of the reasoning of the solution to problem 122, taking that $a = g$.

124. If the ball begins to slip from height h (Fig. 65), it will reach the bottom with a velocity $v_0 = \sqrt{2gh}$. Its further motion upwards will be uniformly retarded and have a velocity $v = v_0 - at$, where a is the acceleration imparted to the ball by the force of gravity.

For the ball's motion on the right-hand inclined plane

$$a = g \sin \beta,$$

therefore

$$v = v_0 - gt \sin \beta.$$

Obviously the ball will move upwards along the inclined plane until its velocity v equals zero, i.e. after time

$$t_1 = \frac{v_0}{g \sin \beta}.$$

The ball will move downwards for the same period of time, and so the full time of its motion on the right-hand inclined plane will be

SOLUTIONS

$$T_1 = 2t_1 = \frac{2v_0}{g \sin \beta}.$$

Similarly for the left-hand plane we have:

$$T_2 = \frac{2v_0}{g \sin \alpha}.$$

The full period of oscillation of the ball will be

$$T = T_1 + T_2 = \frac{2v_0}{g} \left(\frac{1}{\sin \alpha} + \frac{1}{\sin \beta} \right).$$

Substituting for v_0, we shall obtain:

$$T = 2 \sqrt{\frac{2h}{g}} \left(\frac{1}{\sin \alpha} + \frac{1}{\sin \beta} \right)$$

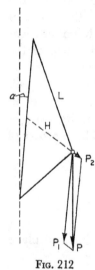

125. Resolve the weight P of the attached load into two forces (Fig. 212) : P_1, acting in a line parallel to the inclined axis, and P_2, acting perpendicularly to P_1. The latter force equals $P \sin \alpha$. This force will act in a similar way to that in which the force of gravity acts on an ordinary pendulum, but the force parallel to the axis will not exert any influence on the pendulum's oscillations. H, the height of the triangle, serves as the length of the mathematical pendulum given. In this case it will equal

$$L\frac{\sqrt{3}}{2}.$$

Therefore the period of small oscillations of the pendulum will be

FIG. 212

$$T = 2\pi \sqrt{\frac{L\sqrt{3}}{2g \sin \alpha}}.$$

126. (a) When the charge is placed below the pendulum's bob, in a vertical line with the point of suspension, the ball will be subjected to the force of attraction to the positive charge, according to Coulomb's Law, as well as to the force of gravity. This force arising from Coulomb's interaction will give a component acting towards the position of equilibrium and thereby increasing the restoring force. Thus oscillation will take place as if the acceleration due to the force of gravity g were increased. Therefore the pendulum's period of oscillation will be reduced.

(b) When a positive charge is placed at the point of suspension O, the force of Coulomb's interaction of charges of opposite signs will always be acting in the line of the thread, therefore the restoring force will be unchanged and the ball's period of oscillation will remain the same as it was.

(c) When the charge is placed to one side on a level with the lowest position of the ball, the force of interaction between the charges during one half-oscillation (to the right of the position of equilibrium) will be opposed to the restoring force and therefore will decrease the pendulum's acceleration; during the other half-oscillation (to the left of the position of equilibrium) it will act in the same direction as the restoring force and will therefore increase the pendulum's acceleration. But the force of interaction decreases as the distance increases and will be on average greater when the bob swings on the side nearer the charge (i.e. when the charge is decreasing the bob's acceleration) than on the side further from the charge (i.e. when the charge is increasing the bob's acceleration). Therefore a charge placed to one side will, on balance decrease the pendulum's acceleration and consequently increase the period of oscillation. Besides, the centre position about which oscillation occurs will be shifted towards the charge.

127. Since radio-waves travel at a velocity of 300,000 km/sec, the receiving sets receive the signals almost instantaneously and we may consider that the signals from the transmitter at A reach B and C at the same moment, although they are located at different distances from A. But the sound travels from C to B with the velocity of sound (330 m/sec). Therefore the distance from B to C will be the distance travelled by sound in 1 sec, i.e. $BC = 330 \times 1 = 330$ m.

128. The frequency of beats equals the difference between the frequencies of the string's vibrations and those of the tuning-fork. Attaching a weight to the tuning-fork will decrease the frequency of its vibrations, while tautening the string will increase its vibration-frequency.

Therefore if the attaching of a weight reduces the frequency of the beats, this means that the tuning-fork's frequency has approached the string's, which is less than that of the tuning-fork. To tune the string to resonance with the tuning-fork the string should be tautened.

129. The pitch (frequency of vibration) of a reflected sound does not equal that of the original sound in cases when the source

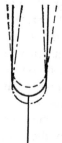

of sound, or the obstacle from which the sound is reflected, is moving (Doppler effect).

130. The prongs of a tuning-fork working normally move in opposite phase, i.e. they are always moving in opposite directions (Fig. 213). Therefore the centre of gravity of the tuning-fork remains stationary and consequently no external force is required to cause these oscillations. The tuning-fork can make its oscillations without being rigidly fixed.

Fig. 213

If one of the prongs be cut off, and the remaining prong makes oscillations of the same sort as before, then the centre of gravity will no longer remain stationary. Consequently an external force must act in order that these oscillations should occur, i.e. the tuning-fork must be rigidly fixed (e.g. the handle should be clamped in a vice); there is then an outside force acting, on the part of the clamp, which brings the centre of gravity into motion. But if the handle be simply held in the hand, the fork will not be sufficiently rigidly fixed and oscillations of the previous type cannot occur.

Thus the existence of two prongs makes it unnecessary to clamp the tuning-fork rigidly, i.e. it allows the instrument to be used when the handle is held in the hand.

131. By the second law of dynamics the impulse of a force applied to a given body equals the change in the body's momentum. Denoting the force acting on the pig by F, and the pig's velocity by v, we shall have:

$$F \times 0{\cdot}01 \text{ dyn/sec} = 5000\, v \text{ gcm/sec.}$$

For the pig to overtake its own squeal, it must move with a velocity greater than the velocity of sound, i.e. v must be greater than 330 m/sec. Therefore

$$F > \frac{5000 \times 33{,}000}{0{\cdot}01} \text{ dyn}$$

or

$$F > \frac{5000 \times 33{,}000}{0{\cdot}01 \times 980 \times 1000} \text{ kg wt.}$$

$$F > 16{,}500 \text{ kg wt. approx.}$$

132. A man hears a sound when variable pressure acts on his ear-drum. At the antinode of displacement of a stationary wave the amplitude of the oscillations has its maximum value, while the amplitude of pressure is practically equal to nought. But at the node of displacement of a stationary wave, the amplitude of displacement equals nought, while the pressure oscillates with maximum amplitude. Therefore a man should hear a louder noise at the node of a stationary wave.

133. In an empty hall the sound is reflected from the walls almost without any absorption taking place, but the clothing of the people in the hall is a material which absorbs sound.

VIII. The Mechanics of Liquids and Gases

134. Since the float is in equilibrium on the surface of the vessel, its weight is equal to the weight of the liquid which its displaces. Therefore, if the float were to be replaced by the liquid in which it floats, this liquid would occupy a volume equal to the volume of the submerged part of the float, and the level of the liquid would not change. Consequently the law of communicating vessels will not be broken if there is a float on the surface of the liquid in one of the vessels.

135. The pressure of the water at the level of the lower end of the tube is the same both inside and outside the tube. Consequently the string is being pulled downwards by the weight of the tube itself, less the force of upthrust (the weight of the same volume of water as the tube's capacity), and also the pressure which the water exerts on the closed end of the tube (atmospheric pressure enters into this). And the pressure of the column of water is proportional to the depth of immersion of the closed end of the tube.

In fact, if the open end of the tube be imagined to be closed, nothing will be altered. But then the following forces will act on the strings: the weight of the tube and its contents, the pressure of the column of water above the tube plus atmospheric pressure, and lastly the pressure exerted by the water underneath, which is equal to the weight of a column of water whose height equals the depth of the open end of the tube beneath the free surface of the water, plus atmospheric pressure. But this last force is balanced by the water and the piston inside the tube.

136. When the bucket is under water a force must be exerted equal to the weight of the bucket, less the weight of water displaced by the bucket itself. Taking the weight of unit volume of water to be d_0, we find $P-(P/d)d_0 = P(1-d_0/d)$. But when the bucket has been drawn out of the water, a force must be applied equal to the weight of the bucket and the weight of the water in the bucket, i.e. $P + Vd_0$.

137. If a body is floating on the surface of a liquid, then, according to Archimedes' principle, its weight is equal to that of the displaced liquid. Thus, if the body be replaced by an equal weight of the liquid, the level of the liquid will not be altered, and therefore the amount of work required to raise the level of the liquid through a height h will not be altered either.

138. The change in the acceleration caused by the force of gravity which takes place on a change in latitude acts equally on the steamer and on the water in which the steamer floats. Therefore the steamer's draught will not alter.

139. The lifting power of an aerostat equals the difference between the weight of air displaced by the aerostat and the weight of gas it contains. In other words, the lifting power is proportional to the difference between the densities of air and of the gas which fills the aerostat. As the density of a gas is in inverse proportion to absolute temperature, so the difference in densities is also in inverse proportion to absolute temperature, i.e. the lifting power is greater, the lower the temperature.

140. The interpretation given is not correct. When the flask containing air is put on the scales, the weight of flask plus air contained in the flask is being found, *less* the weight of the volume of air which the flask can contain (according to Archimedes' principle), i.e. the weight of the flask alone is found (without air). Then, when the air has been evacuated from the flask, the flask is weighed and this time the weight of the flask alone (without air) is found, *less* the weight of air which the flask can contain. The difference between these two weighings gives the weight of air which the flask contains.

141. When the sac is filled with air at atmospheric pressure, its weight will remain unchanged. In fact the weight of air in the sac is balanced by its lifting power, since the weighing takes place in air and not in a vacuum (we neglect the volume of the material from which the sac is made). If the air in the sac is compressed, i.e. its density is greater than that of the surrounding atmosphere, then the weight of the sac will be increased. But in this case determination of the density of the air demands that not only the volume and weight of the sac be known, but also the pressure inside the sac. If we know how many times the pressure, and so the density, of the air inside the sac is greater than that of the surrounding atmosphere, and take into account the sac's lifting power, we can find the density of the air inside the sac and also that of the air outside it at atmospheric pressure.

142. If the scales are in equilibrium the weights in one pan will equal the combined weights of the glass tube and the column of mercury in the tube which is above the level of the mercury in the vessel beneath.

In fact the closed, upper end of the tube is subject to atmospheric pressure, i.e. to the force exerted by the column of air above, the weight of which equals in practice the weight of the column of mercury in the tube. The thickness of the glass sides need not be taken into account, since atmospheric pressure also acts on them from below. But atmospheric pressure does not act on the tube from below. (Atmospheric pressure causes a rise of the mercury in the tube, but since the mercury does not reach the top end of the tube, this pressure is not transmitted to the top end.)

In this we are neglecting the loss of weight of the end of the barometer tube which is immersed in the vessel.

143. The Torricellian experiment will not work in this case. Although, in a vertical position, there will be a position of equilibrium obtaining for the two liquids, this will be unstable equilibrium, since the common centre of gravity of the liquids does not occupy its lowest position. Therefore the mercury will flow out of the tube and the water will flow in.

144. The water will sink and will not prevent the access of air (which supports combustion) to the kerosene.

145. The water is fed into the jacket from the bottom end so that it should fill the whole jacket. If this were not done, the jacket would be partly filled with air (Fig. 214).

146. This would be impossible if there were air above the surface of the water since the pressure of the air would then have been equal in the first case but not equal in the second, and the water could not have remained on the same level. Therefore the U-tube was evacuated before being soldered.

As to the pressure of water-vapour, part of the water-vapour in one limb is condensed

FIG. 214

when the tube is tilted, while a corresponding amount of water in the other limb turns to vapour, and the pressure of water-vapour above the level of the liquid remains the same in both limbs. Therefore the level of the liquid in each limb remains at the same height.

147. The pressure on the membrane from inside equals the pressure of air at the open end of the tube, less the weight of a column of hydrogen, whose base area equals that of the tube's cross section, and whose height is h (Fig. 215). The pressure on the membrane from outside equals the pressure of air at the open end less the weight of a column air of the same cross section and height h. Since air is heavier than hydrogen, the pressure from out-

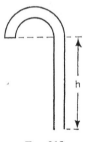

FIG. 215

side is reduced by a greater amount than that inside, i.e. the external pressure is less than the internal pressure. Therefore the membrane will bulge outwards.

148. The pressure on the valve from either side will not equal the pressure registered on the gauges. To the latter pressure must be added the hydrostatic pressure of a column of gas $\rho g h$. Since the density of CO_2 is greater than that of H_2, the pressure on the valve from the left will be greater than from the right, and after the valve is opened part of the carbon dioxide will pass into the vessel containing hydrogen.

In the second experiment the pressure-gauges will be underneath. In this case the pressures at the valve will be $\rho g h$ less than those registered by the gauges. If the pressure-gauges read the same, then the pressure of hydrogen on the valve is greater than that of the carbon dioxide, and when the valve is opened, part of the H_2 will pass into the vessel containing CO_2

149. The purpose of a water-tower is to create hydrostatic pressure in the pipes. The flow of liquid from a tap is governed by the hydrostatic pressure, which equals the difference between the weight of a column of water of base area of 1 cm² and height h and the weight of a similar column of air (Fig. 216).

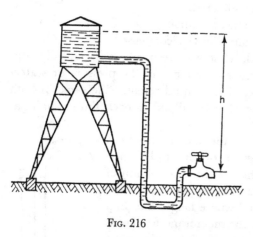

FIG. 216

Let us see whether a gas tower can create pressure in a gas-supply system. Suppose that a tank contains gas under pressure p, which is equal to the pressure of the surrounding atmosphere. Then the pressure at a tap below will equal $p + \rho g h$ (where ρ is the density of the gas), i.e. it will be higher than in the tank. But the pressure of the atmosphere below will equal to $p + \rho_0 g h$ and since the density of air ρ_0 is greater than the density of lighting-gas (assuming that both are subject to approximately the same external pressure) it follows that the pressure of the atmosphere below will be greater than the pressure in the gas-pipes. So a 'gas-tower' cannot by itself cause an increase in the pressure in the gas-pipes.

However, if the gas in the gas-tower is under high pressure, so that its density is greater than that of the surrounding atmosphere, then a 'gas-tower' will produce a certain excess pressure. It is both difficult and dangerous to create a very high pressure in a 'gas-tower' in practice, and the excess pressure that results would be insignificantly small by comparison with the pressure created by a water-tower. Therefore the use of 'gas-towers' is not a practical proposition.

150. The excess pressure below will be greater as a consequence of the fact that a column of compressed air weighs more than a similar column of the air outside the tube. This difference in weights is additional to the excess pressure which exists at the top.

151. If the column of gas were stationary, the excess pressure of the gas over atmospheric pressure ought to be greater on the upper floors than on the lower (see problem 149). But when the gas is moving along the pipes, there is a fall in the pressure, owing to the action of the force of friction. Therefore the gas-pressure decreases as the gas gets farther from the source of supply. This decrease can be so great that the gas-pressure in the upper storeys is less than in the lower ones (since the gas is brought into the house by underground pipes).

152. As the balloons rise, the pressure and density of the surrounding atmosphere decrease. Consequently, if the volume of the balloon remains constant, its lifting power, which equals the weight of air displaced by its volume, decreases as the balloon's altitude increases. Since the rubberized fabric hardly stretches at all, the volume of the balloon made from this material will remain almost constant and its lifting power will decrease with the increase in altitude. When the lifting power falls to a level equal to the weight of the envelope and the hydrogen contained in it, there is no further rise—the balloon reaches its ceiling. On the other hand, the balloon of thin rubber expands easily, and therefore, as the outside pressure decreases with a gain in altitude, it will swell. Thanks to the increase in volume, the balloon's lifting power will not noticeably decrease as it rises (notwithstanding the decrease in the density of the air), and the balloon will continue to rise. Thus a balloon made of thin rubber will rise much higher than a balloon made of rubberized fabric. Pilot-balloons and sond-balloons, made of this thin rubber, are used for meteorological observations and rise to very great heights—up to 30–40 km.

153. The bubble is larger in winter since the thermal coefficient of expansion is greater for the liquid than the glass. The volume of the liquid is reduced more by cooling than the volume of the glass container and the space occupied by vapour increases. In summer, on the other hand, the liquid expands when the temperature rises, and since the vapour is easily compressed under pressure, the volume occupied by the vapour decreases in size.

This can be observed by experiment. Take a flask, fill it with water and bring the water to boiling-point. Part of the water will be driven out of the flask during this. Cool the flask slightly so that it can be corked and so that the water fills it entirely. Lay the flask horizontally. As the flask cools further a bubble will appear above the water and if the cooling process be continued, the dimensions of the bubble will increase.

154. The internal pressure on the cork equals the weight of the column of mercury above the level of the opening and this is less than the pressure of the atmosphere which balances the whole column of mercury. Consequently, if the cork be removed, air will enter the tube and the mercury will fall till its level in the tube is the same as its level in the vessel.

155. It is evident that on a level with the lower end of the tube *d*, which passes through the neck of the vessel, that is, on a level with opening *b*, the pressure of the liquid is equal to that of the atmosphere. Therefore, when opening *a* is uncorked, the pressure outside will be greater than that inside and bubbles of air will enter the vessel, the level of the water in the vessel will fall and it will enter the tube d. When the level of the water in the tube reaches the level of opening a, the air will cease to enter the vessel. If opening b be uncorked instead of a, equilibrium will be maintained as a result of the equality of external and internal pressures, i.e. air will not enter the vessel, nor will water pour out of it. If opening c be unstopped, then the water will pour out of the vessel and air will enter through tube d. However, the pressure at the lower end of the tube will remain equal to that of the atmosphere, regardless of changes in the level of water in the vessel, and water will pour out at a constant velocity until the level of the water falls to the level of the lower end of tube d. After that, the water will continue to pour out, but the speed of flow will decrease.

156. Changes of pressure in the limbs of the pressure-gauge are related to changes in the volume of air inside the apparatus by Boyle's law.

So we have:

(1) Tap K open, no powder. The level of the mercury is brought to the upper mark on the bulb B. The air in flask A occupies a certain volume v. The pressure-gauge registers atmospheric pressure H.

(2) Tap K closed, no powder. The level of the mercury is brought to the lower mark on bulb B. The air occupies a volume $V + v$. Pressure is $H-h$.

(3) Tap K open again, powder has been poured into flask A; when the mercury level has been brought to the upper line on the bulb, i.e. at atmospheric pressure H, the tap K is closed. Evidently the air now in the flask and the connecting apparatus occupies a volume v', such that $v-v' = v_1$, the volume of the powder.

(4) Tap K closed. The mercury level is brought to the lower line on the bulb, as a result of which the volume of air becomes equal to $v' + V$, and the pressure changes to h'.

From Boyle's law we have:

from (1) and (2)
$$vH = (v + V)(H-h),$$
$$v = \frac{H-h}{h} V.$$

From (3) and (4)
$$(v' + V)(H-h') = Hv',$$
$$v' = \frac{H-h'}{h'} V.$$

Hence the volume of the powder equals
$$v-v' = \frac{H-h}{h} V - \frac{H-h'}{h'} V = \frac{VH(h'-h)}{hh'}.$$

157. When the box is immersed in the water, the air inside it will be compressed and water will enter the box. The volume of air in the box can be found from Boyle's law:
$$h_0 S P_0 = h_1 S P_1, \tag{1}$$
where, for the respective positions of the box (not immersed and immersed), h_0 and h_1 represent the height of the lid of the box above the level of water in it, P_0 and P_1 represent the pressure of air in the box and S is the area of the base. It is more convenient in the given instance to express pressure in terms of metres of the column of water. Substituting $h_0 = 3$, $P_0 = 0.76 \times 13.6$ (metres of the column of water), $P_1 = P_0 + 18.6 + (3-h_1) = 10.3 + 18.6 + 3-h_1$ (metres of the column of water), we obtain from equation (1) a quadratic equation for h_1:
$$h_1 = \frac{3 \times 10.3}{31.9 - h_1};$$
solving this, we get $h_1 = 1$ m approx.

Thus the new (compressed) volume of air in the box equals approximately 1 m^3. Neglecting the volume occupied by the sides of the box, we find that the upthrust equals the weight of water in a volume of 1 m^3, i.e. 1 tonne wt.

158. It is simplest of all to attach a sinker with a specific gravity of more than one, to the body and so cause its immersion in the water. The weighing of body and sinker must be carried out first in air and then in water. The specific gravity of the weight must be found first, which can be done by the normal method. Knowing the weight of the sinker and its specific gravity, it is possible to find the specific gravity of the body being tested with the help of the following calculation.

Let the weight of the body being tested be P_1, that of the sinker be P_2 and its specific gravity d_2. The weight of the body under test plus the sinker in water is P'. The difference between the weight of body and sinker in air and in water allows us to find the volume of the body plus sinker

$$v = \frac{P_1 + P_2 - P'}{d}$$

(d being the specific gravity of water).

Since v equals the sum of the volumes of the body under experiment v_1 and of the sinker v_2 and since

$$v_2 = \frac{P_2}{d_2},$$

it follows that

$$v_1 = v - \frac{P_2}{d_2}$$

and the unknown specific gravity

$$d_1 = \frac{P_1}{v_1} = \frac{P_1}{v - \dfrac{P_2}{d_2}} = \frac{P_1 d d_2}{(P_1 + P_2 - P)d_2 - P_2 d}.$$

In the Physics Olympiad of Moscow State University in 1939, in which this problem was set, several schoolchildren gave another solution which is also possible, though more difficult to realize. A diagram of this solution is given in Fig. 217. A hook is set in the bottom of the vessel with a pulley attached; a thread is passed through the pulley and attached to the body at one end and the scales at the other. Then weights must be placed on the other scale-pan to achieve equilibrium; these weights will be equivalent to the upthrust, less the weight of the body.

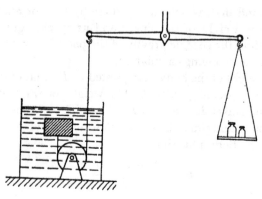

FIG. 217

159. While the bucket is falling freely, the upper layers of water will cease to exert pressure on the lower ones and therefore there will be no pressure on the side-walls of the bucket. Thus the water will cease to flow out.

160. The initial difference between the levels of water in the test-tube and the vessel can be found from the equation

$$P_0 = P - \rho g h,$$

where P_0 is the external pressure, P is the pressure of air inside the test-tube and ρ is the density of the water. A freely falling body behaves as if it had no weight. Therefore, when it is falling freely, the column of water will not exert hydrostatic pressure and consequently the level of water in the test-tube will fall until $P = P_0$ (Fig. 218).

161. The surrounding water exerts pressure on a body immersed in it. This pressure acts on both the upper and lower

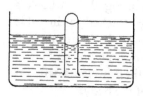

FIG. 218

parts of the body. But since pressure increases with depth, the forces acting on the lower part of the body upwards are greater than the forces acting downwards on the upper part of the body. It is the difference between these two forces that determines the upthrust. When a

submarine is pressed close to mud so that there is no water between the submarine and the sea-bed, there is no pressure exerted
by the water on the
lower part of the vessel,
i.e. there is no force acting upwards. But the
pressure exerted on the
upper part of the sub-
marine continues to act
downwards and, to-
gether with the sub-
marine's weight, presses
the vessel to the sea-
bed (Fig. 219).

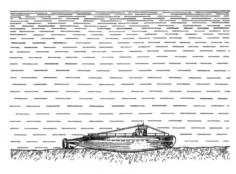

FIG. 219

162. Above the water in the flask will be saturated water-
vapour at the temperature at which water boils at the given
atmospheric pressure. It is a condition of boiling that the temperature of the saturated vapour and the outside pressure should be
equal. Therefore the pressure at the
end of the manometer connected to
the flask will always be equal to the
outside pressure, and consequently
the manometer will always register
zero, regardless of the height above
sea-level.

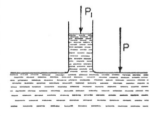

163. A suction pump raises water
at the expense of the difference

FIG. 220

between the outside atmospheric pressure $P = 1$ atm and the
residual pressure in the pump P_1 (Fig. 220). The height h through
which the water is raised is to be found from the formula

$$\rho g h = P - P_1,$$

where ρ is the density of the water, and g the acceleration due to
gravity. Hence we find that

$$h = \frac{P - P_1}{\rho g}.$$

In the case of hot water ($+ 90°C$), the residual pressure P_1 in the pump will be increased at the expense of the saturated vapour pressure of water vapour at $+ 90°C$ (which is approximately equivalent to 53 cm of mercury). Therefore $(P-P_1)$ when cold $> (P-P_1)$ when hot. Consequently the hot water will be raised a considerably smaller height. It is true that the density of hot water is somewhat less than that of cold, but this change is much less than the change in the saturated vapour pressure of the vapour; therefore this last factor does not alter the principle of the answer.

164. According to Archimedes' principle the weight of floating ice equals the weight of water displaced by it. Therefore the volume of water which will be formed as the ice melts will be exactly equal to the volume of displaced water and the level of water in the beaker will remain unaltered.

If the beaker contains a liquid denser than water, then as the ice melts the water so formed will have a greater volume than that of the liquid displaced by the ice, and the water will over-flow. The reverse will obtain if the liquid is less dense than water and the level will fall.

165. Since the piece of ice containing the lead weighs more than a piece of pure ice of the same volume would do, it sinks lower in the water than the piece of pure ice and displaces a greater volume of water than will be formed as the ice melts (see previous problem). Therefore when the ice melts the level of the water will fall (the piece of lead will then sink to the bottom, but its volume remains the same and it does not directly alter the level of the water).

If there are bubbles of air in the ice, the piece of ice will weigh less than a piece of solid ice of the same volume and consequently it will not sink so deep as a solid piece of ice of the same volume. However, since the weight of the bubble of air can be ignored (by comparison with that of the ice), the piece of ice displaces, as before, the same weight of water as its own weight; when the ice, melts, the level of the liquid will not alter (when the ice melts, any bubbles will rise to the top and leave the water). Therefore the case when a bubble is contained in the ice is not the reverse of the case when lead is contained.

166. According to Archimedes' principle an upthrust is exerted by a liquid on a body immersed in the liquid, the upthrust being equal to the weight of liquid of the same volume as that occupied by the immersed body. This force reduces the tension in the thread from which the body is suspended. Therefore the following forces are acting on the right-hand scale-pan: the weight of the stand, and the weight of the body less the weight of the displaced water. According to Newton's third law, a body immersed in liquid will exert a force on the liquid equal to the upthrust, and this will be transmitted by way of the liquid in the vessel to the left-hand scale-pan. The additional force indicated comes into play because the level of water is raised by the immersion and consequently the pressure on the bottom of the vessel is increased. So the following forces will be acting on the lefthand scale-pan: the weight of the vessel and water, and the weight of a volume of water equal to the volume of the immersed body. Since the stand plus body weighs the same as the vessel plus water, equilibrium can be restored by placing on the scale-pan which supports the stand a load equal to twice the weight of a volume of water equal to the volume occupied by the body.

167. The water exerts an upthrust on your finger when it is immersed. A force opposite and equal to this upthrust, according to Newton's third law, will act on the bottom of the vessel (see problem 166). Equilibrium will be destroyed and the scale-pan on which the vessel stands will sink.

168. At a temperature of 0° the force acting on a body immersed in a liquid equals $P - v_0 d_0$, where P is the weight of the body, and $v_0 d_0$ is the weight of a volume of the liquid equal to the volume occupied by the body (v_0 being the volume of the body, d_0 being the specific gravity of the liquid). As the temperature is raised, the volume of the body increases, while the specific gravity of the liquid decreases. The weight of the same volume of liquid as that of the immersed body at temperature t equals $\frac{v_0(1 + \alpha t)d_0}{1 + \beta t}$. Since the thermal coefficient of cubical expansion for solid bodies (α) is usually less than that for liquids (β), it follows that $\frac{v_0(1 + \alpha t)d_0}{1 + \beta t} < v_0 d_0$. Therefore in the majority of cases the upthrust exerted on the body by the liquid decreases with a rise in temperature, i.e. the scales incline to the left (Fig. 87).

169. Rotation of the shaft does not take place since the water exerts pressure upon the curved surface at all points in a direction perpendicular to the surface, i.e. in the line of the radius produced. Since these forces pass through the centre of the shaft, they cannot cause the shaft to rotate. All these forces have a resultant, which acts outwards and in a slightly upwards direction, but which passes through the axis of the shaft and which therefore only tends to push the shaft out of the side of the vessel and not to make it rotate about its own axis.

170. When the beaker is dipped into the water, the force which must be applied to it to achieve this will increase with the depth to which the beaker is dipped, since the upthrust will increase. However this force will not be the same in the two cases. In the first case the air inside the beaker is compressed and the water partly enters the beaker. Thus if the beaker be dipped to the same depth on both occasions, the volume of water displaced will be less in the first case, so the thrust will also be less than in the second case: therefore the force equalling this upthrust that must be applied to dip the beaker in the water is less in the first case, and so is the work done by this force in immersing the beaker to the same depth.

171. A certain amount of water remains inside the U-shaped part of the siphon-tube. This water acts as a cork, shutting the gases from the drainage system off from the house.

172. This experiment is not valid since the level of the liquid in the capillary tube falls for two reasons: (1) as a result of atmospheric pressure on the liquid and (2) as a result of the expansion of the vessel, which is due again to the atmospheric pressure, which previously acted only on the outside, but which now also acts on the inside. These two causes are not distinguished in the experiment described.

173. The press can work, since the principle underlying its operation is valid for gases as well as for liquids (that pressure is transmitted in all directions in a gas, just as in a liquid); but the coefficient of efficiency will be extremely small since the preponderant part of the work done will be expended on compressing the gas, since the compressability of gas is great—considerably greater than that of liquids.

174. The destruction caused by an explosion is determined by the work which the steam or the liquid can do in expanding from its initial volume to the volume which it should occupy at atmospheric pressure. And this work depends as much on the change in volume as on the amount of pressure. Steam, like gas, has great compressability, while liquids have very little; therefore compressed steam, even at a comparatively low pressure (e.g. 15 atm) can do much more work in expanding than a liquid can, even though the latter is at a very much higher pressure (e.g. 600 atm), for the increase in the volume of the gas while its pressure falls to 1 atm is incomparably greater than the liquid's increase during its fall of pressure. In fact, if steam is under a pressure of 15 atm, then its volume will increase 15-fold in reaching atmospheric pressure. But water which starts at a pressure of 600 atm increases its volume by only 0·03 of its initial volume in reaching atmospheric pressure.

175. Factory chimneys are built very tall in order to increase the draught in the furnaces. At the upper end of the chimney, the pressure of the gases leaving the chimney is equal to atmospheric pressure at that height. Therefore the pressure of the air at the lower end of the chimney must be greater than the pressure of the hot (and therefore lighter) gases inside the chimney. This excess of external pressure is what creates a draught in the furnace and so causes the movement of the hot gases up the chimney.

The greater the column of gas, i.e. the taller the chimney, and the greater the temperature of the gases, the greater will be the excess pressure and so the greater will be the draught. Therefore brick chimneys are best, since the heated gases give up less heat through the brick walls of such a chimney to the colder outside air, than through steel ones.

176. The stream of steam which rushes up through the pipe takes with it the smoke and gases from the furnace and increases the draught there.

177. Unlike a Segner's wheel, the tube will not revolve when water flows out of it. In fact the law of conservation of the moment of momentum tells us that in a system which initially has no rotation and to which no moment of force is applied from outside, the overall moment of momentum must remain equal to zero.

In the case of an ordinary Segner's wheel the water which flows out of the bent tubes has a certain moment of momentum. Therefore the vessel of water begins to rotate in the opposite direction, since the sum of the moments of momentum will remain equal to zero. In the case of the inverted Segner's wheel which we are considering, the water which flows out of the straight pipe below has no moment of momentum relative to the axis of rotation of the pipe (since the water is flowing along that axis) and consequently the pipe should not rotate.

It may seem that the liquid as it flows into the bent tubes exerts pressure on the walls of the tubes and so should cause a rotation of the pipe. But besides the pressure, exerted by the water flowing in, on the inside walls of the tubes, there is also a pressure exerted by the surrounding liquid on the outside walls. This pressure is greater than at the openings of the tubes (where the pressure is reduced), and it in fact compensates the turning moment created by the liquid flowing into the tubes.

178. The conditions of stability for a body floating on the surface of water and for a body totally immersed in water are different. Let us consider these conditions for a ship.

In both cases the force of gravity P acts on the ship, applied at the centre of gravity of the ship (c.g.) and an upthrust F (the resultant of the forces of pressure exerted by the liquid), applied at the centre of gravity of the displaced volume of water—or centre of pressure (c.p.). For the ship to be stable it is necessary for these two forces to create a moment, when the ship heels over, which restores the ship to a position of equilibrium. For a ship floating on the surface of the water, this condition will be fulfilled if point M—the point of intersection of the line in which the upthrust acts with the ship's plane of symmetry (this point is called the 'metacentre')—lies higher than the centre of gravity

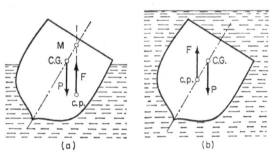

FIG. 221

of the ship. As you see from Fig. 221a, in this case forces P and F will create a moment which restores the ship to its position of equilibrium. Consequently a ship floating on the surface can be stable even if its centre of gravity lies higher than the centre of gravity of the displaced water. This is achieved by the choice of a suitable sectional shape for the ship, so that, when the ship heels over, the centre of pressure shifts in that direction. (The position of the metacentre practically remains unchanged when this happens, so long as the list is not very great.) But if the ship is entirely immersed in water (Fig. 221b) then the centre of pressure evidently lies in the ship's plane of symmetry (coinciding, as it does, with the centre of gravity of the volume of displaced water). And if the centre of gravity of the ship lies higher than the centre of pressure, then the moments of forces P and F turn the ship still farther from its position of equilibrium and the vessel overturns. Thus, so long as the ship is floating on the surface of the water, stability requires that the ship's centre of gravity should lie below the metacentre, which is normally located near the upper edge of the ship's section; but when the ship is entirely below water, stability requires that the centre of gravity should lie below the centre of pressure, which lies approximately in the centre of the ship's section. If the first of these conditions is fulfilled but the second is not, then when the ship is totally below water it overturns.

For the sake of simplification we have taken it that the ship's centre of gravity does not change its position during a list. But if the ship is sinking, this means that it has shipped a lot of water and the position of its centre of gravity depends on where this water is inside the ship. When the ship heels, the water in the ship will flow towards the list, and so the ship's centre of gravity will shift in the same direction. It is easy to see that this tends to increase the ship's liability to overturn still more.

179. When the metal ball is moved from position 1 to position 2 in the liquid, work must be done equal to the change in potential energy of the system, i.e. $mgh - v\rho gh$ or $(m - v\rho)gh$. Since the distance between 1 and 2 is h, the force which must be applied to raise the ball is $(m - v\rho)g$. This force is less than the weight of the ball by an amount $v\rho g$, i.e. by the weight of a volume of liquid equal to the volume of the body. Thus we have reached the usual formulation of Archimedes' principle.

180. The liquid in a rotating vessel also takes part in the rotation. Since outside forces are not acting on the system of vessel and liquid, it follows that the moment of forces equals zero. Therefore the moment of momentum of the system, which is equal to the product of the moment of inertia and the angular velocity, must remain unchanged. When the cork at A is opened, the water will begin to pour out, as a result of which the system's moment of inertia will decrease and therefore the vessel's angular velocity will increase. From the moment when the water ceases to pour out of the vessel from the opening, the speed of rotation will become constant. (As a result of the vessel's rotation the levels of the water in the two parts will not remain horizontal—see problem 183.)

181. Consider a very small volume of liquid of mass m at the surface. The adjacent layers of the liquid must exert upon this volume forces which are normal to its surface (since the liquid moves as a whole [Fig. 222]). If we make this volume a thin layer, then the forces acting on its side edges will be infinitely small and the resultant force N will be normal to the surface of the volume we have isolated for consideration. Besides the force N, its own

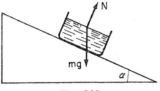

Fig. 222

weight mg will also act upon the volume. These two forces should impart to the isolated volume of liquid an acceleration equal to the acceleration of the vessel as it slips down the inclined plane, i.e. $g \sin \alpha$, where α is the angle made by the inclined plane with the horizontal. Thus the resultant of forces N and mg must be equal to $mg \sin \alpha$. But the force $mg \sin \alpha$ equals the projection of mg on the line of the inclined plane. Consequently the projection of force N on to the same line must equal zero, i.e. the force N is at right angles to the inclined plane, and accordingly the surface of the liquid is parallel to the plane.

182. Consider in isolation a very small volume A of liquid of mass m at the surface of the liquid (Fig. 223). The motion mentioned will cause this volume to move, as does the whole

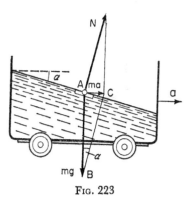

Fig. 223

trolley, with acceleration a. The forces which act on a volume of mass m must produce a resultant equal to ma acting in the same direction as the acceleration a. These forces will be : (1) the weight mg, and (2) the reaction N of the lower levels, normal to the surface of the liquid. The force N must have a value and a direction such that the resultant of N and mg should equal ma. Then in triangle ABC we shall have the relationship $ma = mg \tan \alpha$, whence

$$\tan \alpha = \frac{a}{g}.$$

Consequently, the surface of the liquid, which is perpendicular to the force of reaction N, must make an angle α with the horizontal, in accordance with the above relationship.

183. We know that the surface of a liquid contained in a vessel which is rotating uniformly about a vertical axis passing through the vessel's centre takes the shape of a paraboloid of rotation (Fig. 224a). This is explained by the fact that it is essential to the liquid's rotation that a force acting towards the axis should act on each particle of the liquid, giving it the required centripetal acceleration, which equals the product of the square of the vessel's angular velocity and the distance of the given particle of liquid from the axis of rotation.

For such a force to exist acting towards the axis, the pressure in the liquid must increase from the axis of rotation towards the side of the vessel. Since there is no acceleration in a vertical direction,

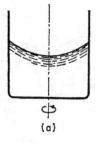
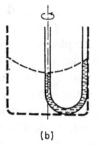

(a)   (b)

FIG. 224

this pressure in the liquid must be equal to the weight of a unit column of liquid from a given depth to the free surface. Consequently the level of the liquid must rise, like the pressure in the liquid, from the axis towards the vessel's side.

To answer the question posed in the problem, let us imagine that the column of water which fills the tube is a part of the water in a rotating vessel. It is directly evident from Fig. 224b that the level of water in the limb through which passes the axis of rotation will fall, while that in the other limb will rise.

184. When there are no springs, the acceleration of both vessel and water is governed only by the earth's attraction, and therefore it is the same for both in every position and during oscillation the water moves along with the vessel, i.e. the surface of the water remains flat and motionless relative to the vessel (Fig. 225a). Also, when the vessel passes through the central position, the surface of the liquid is horizontal, since the vessel has no acceleration in this position, and the surface of the liquid must be perpendicular to the line of action of the force of gravity acting on the liquid.

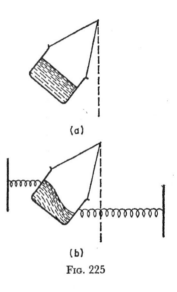

(a)

(b)

Fig. 225

Soft springs attached to the vessel somewhat alter the vessel's acceleration, but do not affect the motion of the water, whose acceleration will depend, as befoie, only on the earth's attraction. In this case, the surface of the water will not remain motionless, but it will oscillate in relation to the vessel (Fig. 225b). As it were a wave passes over the water.

185. The rubber bulb A is made of fairly thick rubber and has two valves at N and M, which open in the same direction, indicated by the arrow in Fig. 226. Another bulb B, made of thin rubber, is connected to it and this in turn is connected to a long

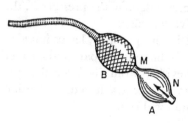

rubber tube which is attached to the atomizer. When the bulb is pressed in the hand, the air inside it closes valve N, opens valve M and passes into bulb B. When bulb A is released, it returns to its original form, thanks to the elasticity of its thick walls. The pressure in

Fig. 226

this bulb falls, and the air outside opens valve N and enters the bulb. At the same time, the air in bulb B closes valve M and part of it enters the atomizer. After the air has been 'pumped' several times into B in this way, this bulb swells considerably (since the air passes out through the atomizer slowly: this is why the bulb is made of thin rubber) and a higher pressure is created inside it, which remains more or less constant. Thus, thanks to bulb B, which acts as a buffer, the flow of air from the atomizer maintains an approximately constant velocity. The bulb B is enclosed in a net to prevent it bursting.

186. When the teaspoon causes the tea in a glass to rotate, a distribution of pressures at the bottom of the glass is set up which increases from the centre of the bottom towards the sides (since the level of tea is higher at the sides than in the middle). After the teaspoon is removed, the pressure at the bottom will be gradually evened out. This leads to the formation of currents from sides to centre and it is these which gather the tea-leaves to the middle of the glass's bottom.

187. A mass of water m, on entering the horizontal part of the pipe has a velocity v relative to the pipe as a result of the engine's motion and consequently it has a kinetic energy of $\frac{1}{2}mv^2$, at the expense of which it can rise through a height h; this height is determined by the condition that all this kinetic energy will be converted into potential energy. Thus

$$\tfrac{1}{2}mv^2 = mgh.$$

Hence

$$v = \sqrt{2gh} = 28 \text{ km/hr approx.}$$

188. Force F_1 equals the weight of a column of liquid of height h and of the same cross-section as the disk S, i.e. $hS\rho g$, where ρ is the density of the liquid. On the other hand force F_2 is determined by the momentum carried away by the stream of water flowing out in unit time, which equals $mv = Sv\rho v = Sv^2\rho$. From Torricelli's formula, v, the velocity of the flow of a liquid out of an opening lying at depth h, equals $\sqrt{2gh}$. Therefore

$$F_2 = 2hS\rho g = 2F_1.$$

The fact that force F_2 is greater than F_1 can be explained by the redistribution of pressure inside the liquid while it is flowing out. When liquid flows out of a wide vessel through a small hole, the lines of flow cluster round the opening and consequently the pressure on the wall of the vessel near the opening decreases, as follows from Bernoulli's law. Therefore the reaction to liquid flowing out is greater than the force of the static pressure on the area of the opening.

189. The water in a water-main is under pressure raised to several atmospheres. As the water flows along the pipe this pressure is gradually reduced as a result of the action of the force of viscosity until it is almost the same as atmospheric pressure—at which it flows out of a fully open tap.

If the tap is covered with a finger, the flow of water in the pipe almost ceases and so the fall in pressure inside the pipe also disappears. Thus the water which passes through the opening still remaining is subject to the full pressure which exists in the water-main, i.e. to a pressure of several atmospheres. The thin stream of water forced out by this high pressure acquires a much greater velocity than does water which is flowing out of a fully open tap.

When the opening is very small indeed, the velocity of the water's flow decreases as a result of the great fall in pressure at the opening itself.

190. It is better to take off into the wind. The lifting power increases with the velocity of the plane relative to the surrounding air. When the plane takes off with the wind following, the plane's velocity relative to the air is equal to its velocity relative to the ground, less the velocity of the wind; but when it takes off into the wind, the plane's velocity equals the sum of these two velocities. Thus a plane taking off into the wind reaches the same airspeed at a lower ground-speed than if it were to take off in a tailwind. Therefore the lifting power reaches the necessary level and the plane leaves the ground at a lower ground speed—which is moie advantageous in many respects, and less dangerous.

191. If the plane simply turned through an angle of 180° about its own longitudinal axis (Fig. 227a), the lifting force which

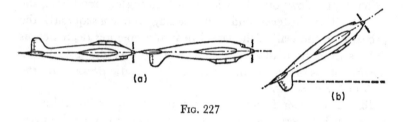

(a)   (b)

Fig. 227

acts upon it—the direction of which depends only on the position of the wings in relation to the oncoming airstream—would be directed downwards and the plane could not only not stay up in the air, but would fall even faster than a freely falling body. To fly upside down, the pilot lowers the plane's tail with the help of the elevator (Fig. 227b), so that the leading edge of the plane's wings should again be above the trailing edge in the plane's inverted position; as a result of this, a lifting force is created which supports the plane.

192. If there were no air-resistance and the planes' engines were not working, the two planes' velocities at C would be the same; though the average speed along arc ABC would be less than that along arc ADC, since velocity increases with the fall

and decreases with the rise of the plane, so that the velocity at
B is less than at D.

In fact, the plane overcomes drag which is proportional to the
square of its velocity, and it is on this that most of the work done
by the engine goes. Since the average speed on arc ADC is greater,
so also is the work done by the engine to overcome frontal resistance greater on that arc. And if the power developed is the same
for both planes, then the kinetic energy of the plane flying along
arc ADC, and therefore its velocity, will be less at C than the
kinetic energy and velocity of the plane flying along arc ABC.

193. It is not possible to blow the filter out of the funnel. The
harder you blow, the more strongly is the filter drawn into the
funnel. This can be explained by Bernoulli's law, according to
which the pressure in a stream of air is lowered where the stream
is narrower. Therefore the pressure is lowered in the narrow gap
between the paper filter and the glass funnel, and the pressure
of the air from outside holds the filter in the funnel.

194. This phenomenon is explained
by Bernoulli's law, according to which
the pressure in a flowing liquid or gas
is least where the velocity is greatest.
By creating a swift stream of air which
passes over the outside leaves of a wad
we reduce the outside pressure on these
leaves and they bend outwards (Fig.
228). This same phenomenon is used
in atomizers, Bunsen burners, Pri-
muses, water-jet pumps and other de-
vices.

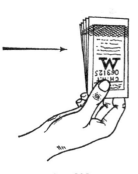

FIG. 228

IX. Heat and Capillary Phenomena

195. The ball passes through the ring at any temperature which is the same for both ball and ring. Heating the ring is equivalent to cooling the ball and consequently the ball will pass freely through the ring when the latter is heated.

196. It is important for a calorimeter that the temperature inside it should even out quickly. This is helped by metal's high thermal conductivity. Therefore the temperature evens out more quickly in a metal container than in a glass one. Besides, the specific thermal capacity of metal is less than that of glass, and this allows the water equivalent of the calorimeter to be reduced and the degree of accuracy of measurements to be increased.

It is also important for a calorimeter that as little heat as possible should escape outside. In this respect too metal calorimeters have an advantage over those made of glass, since metal radiates heat less than glass.

197. The thermos which loses less heat in a given time will be the better one. Exchange of heat takes place between the thermos and the surrounding atmosphere through the sidewalls and the ends of the thermos. If the capacity and the height are the same, the sectional area, and so the end area, is also the same. But the surface area of the side walls is not the same; the surface area of the sides of the cylindrical one is less than that of the square thermos. Therefore the thermos with the circular cross-section

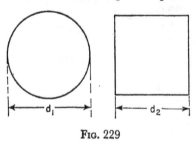

FIG. 229

loses less heat than that with the square cross-section, and it is accordingly better.

For a strict proof of the fact that, given equal sectional areas, the surface area of the side walls of the circular thermos is less than that of the square section, the following calculation may be carried out.

The equality of the end areas (Fig. 229) gives us that

$$\frac{\pi d_1^2}{4} = d_2^2,$$

hence

$$d_2 = d_1 \sqrt{\frac{\pi}{4}},$$

The surface area of the side walls of the cylindrical thermos

$$S_1 = \pi d_1 h.$$

The surface area of the side walls of the square thermos

$$S_2 = 4 d_2 h = 4 h d_1 \sqrt{\frac{\pi}{4}} = d_1 h \sqrt{4\pi}.$$

Comparing these results we find that

$$S_1 < S_2.$$

198. Let t be the final total temperature; let m_1, v_1, t_1 be the mass, volume and temperature respectively of the colder water and let v_1' be the volume of this water at temperature t; also let m_2, v_2, t_2 be the mass, volume and temperature respectively of the warmer water and let v_2' be the volume of this water at temperature t. The equation of thermal balance gives

$$cm_2(t_2 - t) = cm_1 (t - t_1) \tag{1}$$

(c, the thermal capacity, can be cancelled).

On the other hand, the change of the volumes with the change of temperature will be expressed thus:

$$v_1 = \frac{m_1}{\rho_1} = \frac{m_1 (1 + \alpha t_1)}{\rho_0},$$

where ρ_1 is the density of the water at temperature t_1, ρ_0 is its density at a temperature of 0°C, and α is the coefficient of cubic expansion (taking it as constant). Similarly

$$v_1' = \frac{m_1 (1 + \alpha t)}{\rho_0} \; ; \; v_2 = \frac{m_2 (1 + \alpha t)}{\rho_0} \; ; \; v_2' = \frac{m_2 (1 + \alpha t)}{\rho_0},$$

Hence we shall find the alterations in volume :

$$v_2 - v_2' = \frac{m_2 \alpha (t_2 - t)}{\rho_0}, \qquad (2)$$

$$v_1' - v_1 = \frac{m_1 \alpha (t - t_1)}{\rho_0}, \qquad (2')$$

Substituting (1) in (2) and (2'), we find :

$$v_2 - v_2' = v_1' - v_1$$

or

$$v_2 + v_1 = v_2' + v_1',$$

i.e. the total volume of liquid will not be altered.

199. Let m be the mass of the teapot, c the specific thermal capacity of the substance from which the teapot is made, T_1 the temperature to which the teapot is heated when a mass M of water at temperature T is poured into it, and t the temperature of the room. We shall assume that the thermal capacity of the water remains constant and equal to unity throughout. Let us write down the equation of thermal balance :

$$mc(T_1 - t) = M(T - T_1).$$

Hence we find that

$$T_1 = \frac{mct + MT}{M + mc}.$$

Substituting for all these values from the conditions of the problem, we find : for the copper teapot

$$T_1 = \frac{200 \times 0.095 \times 20 + 500 \times 100}{500 + 200 \times 0.095} = 97°C \text{ approx.}$$

for the china teapot

$$T_1 = \frac{300 \times 0.2 \times 20 + 500 \times 100}{500 + 300 \times 0.2} = 91.4°C \text{ approx.}$$

Thus, if there were no surface cooling, the copper teapot would be better for making tea. This result is mainly due to the high thermal capacity of china by comparison with that of copper.

If water from the kettle be poured into the teapot before making the tea, the teapot gets hotter and when water is poured into it the second time (for making the tea), it is hotter. As a result of its higher thermal capacity and its lower thermal conductivity the china teapot, when heated first, will grow cold less quickly than the copper one, and therefore in fact it is better.

200. The feeling of the 'degree' of heat or cold when our bodies touch some object is determined by the amount of heat which is given out, or is received by our bodies in unit time. The thermal conductivity of metal is greater than that of wood. If the metal and the wood are heated to the same temperature, higher than the temperature of our bodies, the metal will transmit to our bodies, on contact with them, more heat in unit time than will the wood. And if the metal is colder than our bodies, it will take from our bodies in unit time more heat than wood at the same temperature will. Therefore in the first case metal will seem hotter than wood, and in the second case it will seem colder. Plainly it is when the metal and the wood are both at the same temperature as our bodies, and when there is no transference of heat, that they will seem equally heated to our touch.

201. The coefficient of thermal expansion of copper is greater than that of iron: therefore when it is heated, the welded plate will bend, as shown in Fig. 230, and this will cause the electric circuit to be broken. As soon as the circuit breaks, heating will cease, the plate will cool and straighten itself out, returning to its previous position and recompleting the circuit; heating will begin again and so on. Such bimetallic plates can consequently be used as inter-

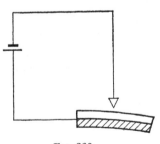

Fig. 230

rupters. Sometimes they are used for automatic shutdown of a section of an electric grid under overloading.

202. When a man's temperature is taken with a clinical thermometer, this is what happens. At first the difference in the temperatures of man and thermometer is considerable and the mercury expands with rapid heating. When the thermometer's temperature is near to that of the man's body, the heating of the thermometer takes place slowly and the mercury also expands slowly. Therefore a considerable time is required for the thermometer to be heated to the temperature of the man. When the thermometer is taken out there is a great difference between the temperatures of the thermometer and the surrounding air, the volume of the mercury contracts rapidly and it is enough to shake the thermometer for the column of mercury to occupy the space which is empty in the bulb.

203. Copper has a high specific thermal capacity and it is thanks to this that when a copper soldering iron is heated, a great amount of heat is transmitted to it. Further, copper has high thermal conductivity, so that a copper soldering iron quickly gives up a large amount of heat to tin or any other material, which must be fused. The other substance which has the same high qualities is silver, but it is too expensive.

204. Consider the state that obtains when the temperature T of the joint is constant. Let c_A and c_B be the thermal capacities of the cylinders, let k_A and k_B be their coefficients of thermal conductivity, and T_A and T_B be the temperatures of their ends. Since the cylinders are homogeneous and heat is not given up through their sides, the temperature in the state that is set up changes along each of the cylinders according to a linear equation and the temperature T of the join can be found from the equality of the flow of heat through any two cross-sections of the two cylinders

$$k_A(T_A - T) = k_B(T - T_B),$$

from which

$$T = \frac{k_A T_A + k_B T_B}{k_A + k_B}. \tag{1}$$

The amount of heat which flows through each cylinder can be found by replacing the variable temperature along the whole cylinder by the average temperature (since the distribution of temperature along the cylinder changes according to a linear equation). The total amount of heat which flows through the cylinders

$$Q = \frac{c_A(T_A + T)}{2} + \frac{c_B(T + T_B)}{2}$$

or

$$Q = \frac{c_A T_A}{2} + \frac{c_B T_B}{2} + \frac{(c_A + c_B)}{2} T. \tag{2}$$

Substituting the value for T from expression (1), we shall get:

$$Q = \frac{c_A T_A}{2} + \frac{c_B T_B}{2} + \frac{(c_A + c_B)(k_A T_A + k_B T_B)}{2(k_A + k_B)}.$$

Bringing the right-hand side over a common denominator and collecting terms containing T_A and T_B, we finally find that:

$$Q = \frac{2k_A c_A + k_A c_B + k_B c_A}{2(k_A + k_B)} T_A + \frac{2k_B c_B + k_A c_B + k_B c_A}{2(k_A + k_B)} T_B.$$

And this expression gives us the answer to the problem posed. Since the coefficients are, in general, different at temperatures T_A and T_B, Q will have a different value if we replace T_A by T_B, i.e. the sum total of heat flowing through the cylinders will in general depend on which of the ends is heated and which is cooled. And since the coefficients at temperatures T_A and T_B differ only in the values of the terms $k_A c_A$ and $k_B c_B$, the amount of heat will be greater if we heat the end of that cylinder for which the product of thermal capacity and thermal conductivity is greater. This is quite clear physically: the role of thermal capacity is obvious; and the greater the thermal conductivity, the less will be the fall of temperature along the cylinders and the higher will be the temperature of all points.

Only in the special case when $k_A c_A = k_B c_B$, as is laid down by our problem, will the amount of heat flowing through the cylinders be independent of which end is heated and which is cooled.

205. Since there are no losses of heat through the sides of the cylinders, the amount of heat which flows through any section of our system in unit time will be the same. If the difference in temperature between any two cross-sections of a homogeneous medium, distance l apart, is $T_1 - T_2$, the amount of heat which flows from the first to the second in unit time through any section between them will be expressed as

$$Q = kS \frac{T_1 - T_2}{l},$$

where k is the coefficient of thermal conductivity and S is the sectional area.

Let T_1 be the temperature of the upper end of the iron cylinder, T_2 be the temperature of the ends in contact, T_3 be the temperature of the lower end of the silver cylinder. Then the amount of heat which flows through any section of the iron cylinder equals

$$Q_{Fe} = k_{Fe} S \frac{T_1 - T_2}{l}.$$

Similarly for any section of the silver cylinder we shall have

$$Q_{Ag} = k_{Ag} S \frac{T_2 - T_3}{l}$$

Comparing these expressions, we shall get:

$$k_{Fe}(T_1 - T_2) = k_{Ag}(T_2 - T_3).$$

Substituting the values of $T_1 = 100°C$, $T_2 = 0°C$ and $k_{Ag} = 11 k_{Fe}$ in the above equation we find that the temperature of the ends in contact is

$$T_2 = \frac{100}{12} = 8 \cdot 3°C.$$

206. Birch wood burns faster than pine wood and therefore more heat is given off from it in unit time.

207. To heat water from 20° C to boiling-point, an amount of heat equal to 80 kcal must be imparted to each kilogramme of water, i.e. work of $80 \times 427 = 34{,}160$ kg must be done. The Dneper hydroelectric station cannot do this amount of work, since 1 kg of water would acquire the energy needed for this only if it fell from a height of more than 34 km.

208. The mica is deformed by the heated metal cone on top of it and a bulge is formed as a result of the heating which is what makes the cone roll. The copper sheet under the mica cools the heated place quickly; thus the cone is in contact with cool mica at every moment and as this cool place is heated it too is deformed. If a sheet of glass lay under the mica, there would be no such cooling of the mica, which would soon grow hot and the cone's movement across it would stop.

209. If the temperature of all the water has already reached 100°C, it will not rise until boiling has ceased, i.e. until all the water has turned to steam. Therefore it is quite useless to turn up the flame. The heat applied should be just enough to maintain boiling in the whole saucepan and to compensate the loss of heat.

210. We know that water has its greatest density at 4°C. If the temperature decreases from a higher temperature to 4°C, the water contracts, but if it cools further from 4°C to freezing-point, i.e. to 0°C, it expands again. When there is a frost, first of all the upper layers of water cool and contract. As they contract, they become heavier than the lower, warmer layers; therefore the layers near the surface

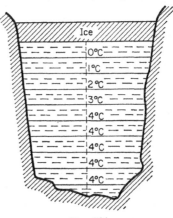

Fig. 231

sink and the deeper, lighter layers rise to take their place. When the whole mass of water has cooled to 4°C, the process of cooling from the surface and the exchange of layers of water stops. At this temperature and below, the upper layers of water no longer sink, because as they cool further they become lighter than the lower layers, i.e. they stay at the surface till they freeze. The top-most layer receives the temperature of the surrounding air, then the temperature gradually rises deeper down, and the layers lying at some depth will have a temperature of $+4°C$ even in the heaviest frost (Fig. 231).

211. Ice melts at $0°C$ only if the necessary amount of heat be imparted to it—about 80 calories to every gramme of ice—while water will only freeze if the same amount of heat is taken from it. Therefore if the vessel is not heated or cooled from outside, the water will not freeze and the ice will not melt. A mixture of water and ice at $0°$ will be in 'thermodynamic equilibrium'.

212. The amount of heat which the water gives up in cooling to $0°C$ is not enough to melt all the ice, since each gramme of water will give up 50 cal in cooling, while heating 1 g of ice to $0°$ and melting it requires that $0.5 \times 40 + 80 = 100$ cal be expended (the thermal capacity of ice is 0.5 cal/deg).

On the other hand, the amount of heat which the ice will take in heating to $0°$ is not enough to freeze all the water, since each gramme of ice will absorb 20 cal in doing this, while for the water to be turned to ice, each gramme of water must be made to give up $50 + 80 = 130$ cal. Thus both water and ice will remain in the calorimeter, and therefore the final temperature of the mixture will be $0°C$.

213. A heater is put underneath because the layers of water that are heated first, being lighter, rise upwards and thus a more effective mixing and heating of all the water is achieved. But when the water is being cooled, the process is exactly the reverse: the colder layers of water, being heavier, sink. Therefore if the cooling agent be put underneath, there will be no stirring effect and cooling will take a very long time. To cool the water more quickly, the ice should be put on top.

214. Water will not boil in a saucepan which is floating in another saucepan containing boiling water, since inevitable loss of heat will mean that the temperature of the water in the floating saucepan is lower than the boiling-point of water.

215. Cooling of a heated body takes place the faster, the greater is the difference between the temperature of the body and that of the surrounding medium. Therefore it is better to allow the water to cool first and then put the ice in.

216. Water is a poor conductor of heat and when the surface is heated by the sun's rays, the heat does not penetrate deep. Besides this, the water evaporates and so is cooled. This often leads to the air, which is heated from below by contact with the heated ground, being at a higher temperature than the water in a reservoir.

217. Yes, if it is put under the mouth of an air-pump and the air be pumped out. Water can boil at room temperature if the pressure of air above it falls to about 15 mm of mercury.

218. The cooling of the lower sphere causes in it intense condensation of vapour. This in turn causes the water in the upper sphere to start evaporating. Evaporation causes the water to cool. As condensation in the lower sphere proceeds faster and faster, so also does evaporation in the upper sphere. As a result of this the temperature of the water in the upper sphere falls so much that the water freezes.

219. In mercury thermometers intended for the measuring of high temperatures, there is a column of nitrogen above the mercury under a pressure of 14 atm; as a result of this, the boiling-point of the mercury is correspondingly raised.

220. For the hydrogen at 0°C we shall have the equation

$$\frac{pv}{273} = \frac{p'v}{283},$$

and for the hydrogen at + 20°C we shall have the equation

$$\frac{pv}{293} = \frac{p''v}{303}.$$

Plainly p'' is less than p', so the column of mercury will shift towards the vessel in which the hydrogen was at first at $+$ 20°C.

221. If a jar is heated non-uniformly, convection-currents are ceaselessly at work in it and at the bottom they act from the cold side towards the warm side. These currents transfer the sediment on the bottom towards the warmer side. In winter the side turned towards the room is plainly warmer than that turned towards the window. So the experiment was performed in winter.

222. Water-vapour will pass from the room outside, since the vapour pressure or density of water-vapour is greater at a humidity of 40 per cent and a temperature of $+$ 20°C, than at a temperature of 0°C and a humidity of 80 per cent.

223. Soap-bubbles of different diameters can not be in equilibrium, because the excess pressure of the surface tension, which acts within the spherical bubble along the line of the radius towards the centre is the greater, the smaller is the radius of the bubble. Therefore the pressure inside the small bubble is

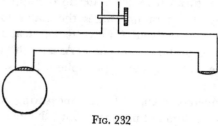

greater and air will flow from it into the large bubble; as a result, the small bubble will contract and the large one will swell. Since tap K is closed, i.e. the volume of air inside the tube and

FIG. 232

the bubbles remains constant, this flow will cease when the radius of curve is the same for both bubbles. As you can see from

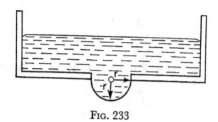

FIG. 233

Fig. 232, this can be when the larger bubble is almost a complete sphere, less a small segment, and the smaller bubble is in the form of this missing segment.

224. Since the liquid does not wet the glass, its surface layer as it curves, exerts a pressure which is directed inwards and equals $\dfrac{2\sigma}{r}$ (Fig. 233), if we consider the radius of the surface layer to be approximately equal to the radius r of the hole. So long as this force is greater than the pressure $\rho g h$, which acts downwards, the liquid will not run out. Thus we have the inequality

$$\frac{2\sigma}{r} > \rho g h,$$

hence

$$h \leqslant \frac{2\sigma}{r \rho g}.$$

225. Since the liquid wets the capillary tube and the tube is pointing upwards, the curve of the meniscus must cut into the liquid, otherwise the liquid could not remain higher in the capillary tube than it does in the wide vessel. The liquid will reach the top of the tube and form a slightly concave meniscus. The forces of surface tension acting at a tangent to the surface of the

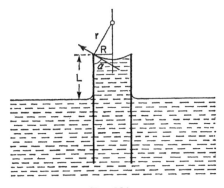

Fig. 234

liquid will form an angle of α with the side of the tube. Therefore the force which supports the column of liquid (Fig. 234) in the tube will be $\sigma \cos \alpha \, 2\pi R$. On the other hand, the weight of the column of liquid in the tube equals $\pi R^2 L \rho g$. Putting the first expression equal to the second, we shall obtain

$$\frac{R}{\cos \alpha} = \frac{2\sigma}{L\rho g},$$

but $R/\cos \alpha$ equals r, the radius of the meniscus of the liquid in the tube.

226. The capillary tube is necessary to obviate errors caused by the surface tension of the mercury in measuring the difference in the level of the mercury, while the wide tube is necessary for the manometer's readings to correspond with the pressure of the gas which is in the vacuum installation.

X. Electricity

227. When the high-tension current is switched on, a charge of static electricity appears on the birds' feathers, as a result of which the feathers spread apart, just as a paper plume opens out when it is connected to an electrostatic machine. This causes the bird to fly away.

228. If a charged conductor be introduced into another, insulated, conductor and brought into contact with the latter's inside wall, the charge will be entirely transferred to the outer surface of this second conductor (Faraday's cylinder).

229. If a ball is surrounded with a concentric metal sphere, an induced charge will appear on the inner surface of the sphere equal in magnitude, but opposite in sign, to the charge of the ball; as a result there will be a charge of the same magnitude and of the same sign as the ball's charge on the external surface of the sphere. This will create the same electric field in the space outside as the ball was creating before. Both the ball and the sphere act in the same way as a charge concentrated at the centre of the ball; therefore the force acting on the piece of paper will remain unaltered.

If the paper is surrounded by the sphere, the force of attraction will be reduced to zero: the paper will be in a Faraday's cylinder, a charge appearing on the sphere, but there being no field inside the sphere. The metal sphere will be attracted to the ball, but not the paper.

230. In the space inside the large sphere, which is charged to a potential of $+ 10,000$ V, all points have the same potential— $+ 10,000$ V. When a small sphere charged to a lower potential is introduced into this large sphere, work is done, as a result of which the small sphere is also charged to a potential of $+ 10,000$ V.

Thus when the smaller sphere is brought into contact with the inside surface of the larger sphere, the charge is transferred, at the same potential, from the inside surface to the outside surface of the sphere, a conductor.

231. Let r be the radius of a small drop and R be the radius of the large drop. Then clearly, when N small drops run together into one big drop, we shall have an equation for the volumes:

$$N. \tfrac{4}{3}\pi r^3 = \tfrac{4}{3}\pi R^3.$$

Hence

$$R = r\sqrt[3]{N},$$

The charge of each small drop equals $q = CV = rV$, since the electric capacitance C of a sphere equals its radius. The total charge of the drops is preserved when they run together to form one drop. Therefore

$$NrV = RV'.$$

Substituting for R, we obtain:

$$NrV = r\sqrt[3]{N}\, V'.$$

Consequently

$$V' = \sqrt[3]{N^2}\, V.$$

232. If the insulating stand connecting A and B begins to conduct, the electrostatic machine is no longer able to maintain a sufficient potential difference between them and there can be no discharge between the spheres. But the subsidiary electrode is well insulated and the potential difference between it and sphere A is close to the potential of the source. Therefore the potential difference is sufficient for a discharge to take place. In addition, the subsidiary electrode-sphere is of small radius and the field at its surface is very strong. Discharge takes place exceedingly quickly, the spark gap becomes conductive for this period, and sphere A acquires the potential of the source's pole; thus a large potential difference is created between spheres A and B which cannot be quickly dissipated since the insulator still conducts electricity only poorly. Therefore, after a spark jumps across the subsidiary spark gap, a spark also jumps across between the spheres.

233. The fallacy in the argument given in the problem lies in the fact that we do not take into consideration the work done in immersing the charges in water and raising them from the water. As the charges approach the dividing line between water and air, polarization charges are caused on the surface of the water so that the motion of both charges in a vertical direction (when the distance between them does not alter) is linked with the performance of work, which cannot be neglected. The work expended on displacing the charges vertically when they are apart will be greater than when they are together (since the field at the edge of the dielectric is greater), and the full amount of work done in the cycle will equal zero.

234. The work goes on increasing the accumulator's energy. Since the plates of the capacitor are connected all the time to the accumulator terminals, the potential difference applied to them remains constant and consequently their charge must decrease. For the capacitor's charge

$$Q = CV = \frac{SV}{4\pi d},$$

where C is the capacitor's capacity, V is the potential difference, S is the area of the capacitor plates and d is the distance between them.

If V remains constant and d increases, Q must decrease. A partial discharge of the capacitor takes place and current flows through the circuit in the direction shown by the arrow in Fig. 235, as a result of which the accumulator is charged.

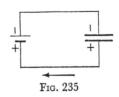

FIG. 235

When the capacitor plates are moved apart, the energy W of the electric field in the capacitor decreases. In fact $W = \frac{1}{2}CV^2$ and if V is constant and C decreases, then W also decreases. The energy which is released in the capacitor also goes on charging the accumulator. Thus all the work expended on moving the plates apart goes on increasing the accumulator's energy and so too does a part of the energy stored in the capacitor.

235. Since the potential difference of the faces of the plates does not change (remaining equal to the e.m.f. of the source), the field strength between the faces does not change either. But the capacitor's capacitance is increased when it is filled with a dielectric. Additional electric charges will flow from the source to the capacitor in such a quantity that the potential difference on the faces of the plates remains the same throughout the filling of the capacitor with the dielectric. And the potential difference will also be at this level after filling.

236. When a positive charge is introduced into a capacitor, a negative charge is induced on the inside faces of the capacitor plates and it remains there, while a positive charge appears on the outside faces. The positive charge on the earthed plate escapes into the earth. The charge on the other plate also escapes into the earth, passing through the galvanometer. So the galvanometer will show a deflection. But when the positive charge begins to escape from the capacitor, the negative charge will begin to flow into the earth from the plates. The charge flowing from one of the plates will pass through the galvanometer, which will then show a deflection in the opposite direction.

237. If the spheres are sufficiently far apart, we can consider that the presence of one sphere does not affect the charge and potential of the other. If the spheres are charged with equal and opposite charges $+ q$ and $-q$ and they are in the air, the potential of the first sphere $V_1 = q/r$, and its capacity $C = q/V_1 = r$. The potential of the second sphere $V_2 = - q/r$ and its capacity also equals r. The potential difference $V_1 - V_2 = q/r + q/r = 2q/r$, and therefore the capacity of the system of the two spheres $C = q/(V_1 - V_2) = qr/2q = \frac{1}{2}r$, i.e. the capacity of this system is half the capacity of one isolated sphere.

This result will become a little clearer if we follow the transition from the case of the isolated sphere to the case we have just considered of two spheres. We can regard the capacity of the isolated sphere as the mutual capacity of this sphere and a sphere of very great radius surrounding it (the charge on the large sphere being of necessity equal and opposite to that of the small sphere). Let us now put the second small sphere into the large one, bearing a charge equal and opposite to that of the first small sphere.(Thus the total charge on the large sphere will be equal to zero.) Since the surrounding sphere is very large, the small spheres inside can be moved such a distance apart that their respective fields do not affect the field of the other. Then the mutual capacity of the two small spheres can be considered as the capacity resulting from the connecting of two capacities in series: (1) the mutual capacity of the first small sphere and the large sphere and (2) the mutual capacity of the second small sphere and the large sphere(these two capacities are shown in Fig. 236 as though they were con-

densers). As in the case of two capacitors connected in series we have here two plates, one of each capacitor connected to-gether, whose total charge is zero (in our version these two plates have been merged into one big sphere) and two plates which are not connected (one of each capacitor; the two small spheres). But we know that the capacity of two similar capa-citors connected in series is half as great as that of each of the capacitors.

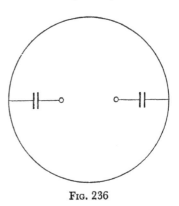

Fig. 236

238. Yes. The leaves of an electroscope always move apart regardless of the sign of the charges on the leaf and the main body, since these charges are always of opposite sign. In the case of alternating current the force of interaction between the leaves and the main body will act in the same direction during both half periods, the leaves will be attracted to the main body and there-fore the average value of this force during one period will not be zero: the leaves will move apart a distance which will depend on the voltage of the circuit.

239. One of the simplest systems is shown in Fig. 237. A second wire is suspended parallel to the overhead wire along the required section of the route; this second wire is connected to the lamp and the lamp's other terminal is connected to the track. Since there is a current passing between the overhead wire and the track, when the tram's arm connects the overhead wire with the second wire, it closes the circuit and the lamp is lit up.

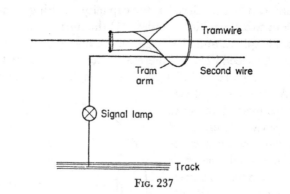

FIG. 237

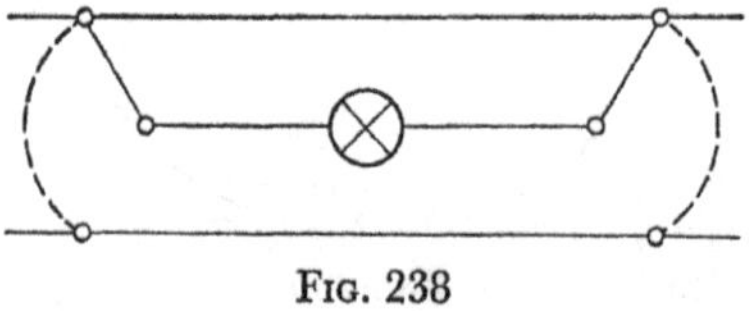

240. Figure 238 represents one of the possible arrangements which satisfy the conditions of the problem. This arrangement requires that instead of ordinary switches, two two-way switches should be put into the system.

FIG. 238

241. If fuse plugs were put on both wires, overloading might cause the fuse to burn out on the neutral wire, but not on the phase wire. Then the light would go out, but the whole wiring system would be under pressure relative to the earth.

242. The carbon filament will incandesce more. The reason for this is that the resistances in carbon and metal filaments depend on temperature in different ways: the resistance of a carbon filament decreases with an increase in temperature, while that of a metal filament increases. At normal incandescence both bulbs have identical resistances (since their power is the same). Therefore when they are connected in series—when incandescence, and consequently the temperature, of both filaments will be lower than normal—the resistance of the carbon filament will be greater than that of the metal filament; therefore the former will incandesce more (its fall of potential will be greater).

243. When appliances R_1 and R_2 are switched on, the fall of potential in the network is increased, the appliances requiring a large amount of current, and the lamps' incandescence is reduced. But the effect of switching on R_2, especially on the incandescence of L_2 is greater than the effect of switching on R_1, since the switching on of R_2 causes a fall of potential in the wires passing through the flat and the cross-section of these wires is less in area, and their resistance accordingly greater, than in the case of the wires which bring the current into the flat.

244. When switch K is closed, i.e. when lamp L_2 is switched on, the resistance in section AB is reduced and so the fall in potential is also reduced in this section. Therefore the strength of the current is reduced in lamp L_2 and increased in lamp L_1.

245. Since the kettle is connected to a fixed voltage V, the amount of heat given off by its heating element in time t is expressed by the formula (from the Joule–Lenz law):

$$Q = 0 \cdot 24 \, \frac{V^2}{R} t,$$

where R is the resistance of the heating element. Since we are neglecting loss of heat to the surrounding atmosphere, we must consider that the same quantity of heat Q is needed in both cases to bring the water in the kettle to boiling-point. If the kettle is to boil, not after 15 min, but after 10, then $15/R_1 = 10/R_2$, i.e. the resistance of the heating element must be reduced $1 \cdot 5$ times. Since a wire's resistance is proportional to its length, the wire of the heating element should be shortened $1 \cdot 5$ times.*

---

* But this may raise the strength of current in the heating element above the permissible level and the element may burn out.

246. There are two circuits in parallel connecting points A and B, each consisting of half one side of the hexagon, a rhombus and then half another side of the hexagon, connected in series. The resistance of each rhombus is R, therefore the resistance of one circuit is $2R$. Therefore the resistance of the whole frame is R. The fact that the vertices of the rhombi are linked at O plays no part in this, since the fact that the two circuits under consideration are identical means that the vertices of both rhombi are at the same potential. Therefore we can calculate the resistances of the two circuits without taking into consideration their link at O.

247. The potential difference at the terminals of the first cell will equal zero if $R = r_1 - r_2$. This can be seen from the following points. The current in the circuit

$$I = \frac{2E}{r_1 + r_2 + R}.$$

The potential difference at the terminals of the first cell

$$V_1 = E - Ir_1 = E - 2\frac{r_1 E}{r_1 + r_2 + R} = \frac{(r_2 - r_1 + R)E}{r_1 + r_2 + R}.$$

Clearly

$$V_1 = 0 \text{ if } r_2 - r_1 + R = 0, \text{ i.e. } R = r_1 - r_2.$$

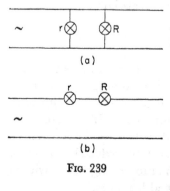

(a)

(b)

Fig. 239

248. If the lamps be connected normally, i.e. in parallel (Fig. 239a), the one which has the lower resistance will shine more brightly. When they are connected in parallel the potential in the two lamps is the same and therefore the power available in the filament of each will be inversely proportional to the resistance, $Q_1 = V^2/r$, $Q_2 = V^2/R$, i.e.

$$Q_1 > Q_2.$$

But if the lamps are connected in series (Fig. 239b), then the potential in the lamps is different, while the current in them is the same. Therefore $Q_1 = I^2r$, $Q_2 = I^2R$, $Q_1 < Q_2$, i.e. the lamp with the greater resistance will burn more brightly.

249. It is possible if $E_2/E_1 < r_2/(r_1 + R)$. If one cell of e.m.f. E_1 and internal resistance r_1 be connected to an external resistance R, then the current passing through the circuit

$$I_1 = \frac{E_1}{r_1 + R}.$$

If a second cell of e.m.f. E_2 and internal resistance r_2 is then added in series, the current

$$I_2 = \frac{E_1 + E_2}{r_1 + r_2 + R}.$$

Clearly $I_2 < I_1$ if $\dfrac{E_1 + E_2}{r_1 + r_2 + R} < \dfrac{E_1}{r_1 + R}$, hence

$$\frac{E_2}{E_1} < \frac{r_2}{r_1 + R}.$$

In other words for this inequality to obtain, the internal resistance of the second cell must be sufficiently great.

250. The voltage between points A and B should equal the e.m.f. of the cell less the fall of potential inside the cell. But the strength of the current in the circuit equals

$$I = \frac{3E}{3r} = \frac{E}{r}$$

where E is the e.m.f. of the battery and r is its internal resistance. Consequently in our example the voltage between points A and B will be $E - Ir = E - E = 0$. Thus a voltmeter connected at points A and B will register zero.

The result we have obtained can be explained in the following way. If the terminals of one battery are short-circuited (i.e. if they are connected by a conductor of infinitely low resistance) the internal fall of potential will equal the battery's e.m.f. and the potential at the battery's terminals will be zero. The circuit under consideration presents a similar picture. Here three identical batteries are connected in series by wires of infinitely low resistance, which is exactly similar to short-circuiting each battery. Therefore the potential at the terminals of each battery equals zero.

Clearly the reading of the voltmeter will not be altered by increasing the number of batteries, if the voltmeter still be connected at points A and B.

251. As you know, a voltmeter connected directly to the source of current shows not the e.m.f. but the potential at the source's terminals

$$V = E - Ir,$$

where r is the internal resistance of the source. Since r is unknown, it is not possible to determine the e.m.f. from the readings of the voltmeter and ammeter with the sliding contact of the rheostat in only one position. But if the sliding contact is moved and the current and potential difference are measured for this new position then we shall obtain the self-evident equation

$$E = V_1 + I_1r = V_2 + I_2r.$$

Hence

$$r = \frac{V_1 - V_2}{I_2 - I_1}$$

and the unknown e.m.f.

$$E = V_1 + I_1r = V_1 + I_1 \frac{V_1 - V_2}{I_2 - I_1}.$$

252. The galvanometer should be connected to the arm to which is normally connected the unknown resistance R_x, and a switch should replace the galvanometer on the bridge's diagonal. Resistances r_1 and r_2 should be selected so that the galvanometer shows the same deflection whether the switch is open or closed. This will mean that there is no current across the bridge's diagonal and consequently the ratio obtains:

$$\frac{R_G}{R} = \frac{r_1}{r_2},$$

therefore

$$R_G = R\frac{r_1}{r_2}$$

253. For connection in series

$$I_1 = \frac{nE}{nr + R} = \frac{E}{r + \dfrac{R}{n}},$$

where E is the e.m.f. of each cell, r is the resistance of each cell, and R is the external resistance.

For connection in parallel

$$I_2 = \frac{E}{\dfrac{r}{n} + R}.$$

Obviously $I_2 = I_1$ if $r = R$, i.e. if the resistance in the wire equals the internal resistance of a cell.

254. If the ammeter shows an absence of current in battery E_2's circuit, then obviously the fall of potential between points c and d caused by the current flowing through battery E_1's circuit is equal in magnitude to the e.m.f. of E_2. And this is what a voltmeter will register if it is placed between c and d.

255. If we connect a voltmeter between points A and B, we find which of the two points has the higher potential. Suppose that we find that the potential at A is higher than that at B. Then we bring the magnetic needle, mounted on a vertical pivot, up under the corresponding wire, e.g. the upper one. The deflection of the magnetic needle's North pole tells us the direction of flow of current through the wire. For example, if the needle's North pole is deflected towards us from the plane of the paper, the current in this wire is flowing through A from right to left. Hence it follows that the source of current in our example is to the right of A.

256. Since the e.m.f. of the battery is opposed to the e.m.f. of the circuit, the resultant e.m.f. acting in the circuit is $115 - 60 \times 1{\cdot}2 = 43$ V. For a current of $2{\cdot}5$ A to flow through the circuit, given this e.m.f., the resistance in the circuit must equal $43 \div 2 \cdot 5 = 17 \cdot 2 \; \Omega$. So the rheostat's resistance must be $17 \cdot 2 - 60 \times 0 \cdot 02 = 16 \; \Omega$.

257. As you know, when the armature of an electric motor rotates, an e.m.f. of induction opposed to the e.m.f. of the source is generated in the wires of the armature. This back e.m.f. increases with the number of revolutions the armature makes per second. The greater the back e.m.f., the weaker is the strength of the current in the armature circuit. If the motor is idling, the armature rotates faster, and when the motor is doing work, the armature rotates more slowly. Therefore, if the motor is loaded a stronger current passes through the armature circuit and therefore a loaded motor heats up more quickly than one that is idling.

258. The amount of heat given off in a circuit with a given resistance is proportional to the average value for the square of the current strength in the circuit. The readings of the hot-wire ammeter are also proportional to this average value (regardless of the kind of current). Therefore if the ammeter reads 5 amp in both cases, then the heating effect will be the same in both cases also.

But an ammeter with permanent magnets (an electromagnetic ammeter) gives readings which are proportional to the average value for the current. It will show deflections, and in the case of pulsating current it will also show the average value of this pulsating current. But the average value of the square of a pulsating current is not equal to the square of the average value of this current. Therefore if an electromagnetic ammeter gives the same readings for direct and for pulsating current, the heating of the furnace will not be the same in the two cases.

259. Let us first consider the influence of change of temperature on ammeter readings.

An ammeter is normally used with a shunt, i.e. a resistance connected in parallel to the instrument. The ammeter's resistance is very small, that of the shunt is even smaller; for this reason, it is usually made of copper, like the ammeter coil. The distribution of current through the instrument's coil and the shunt is determined by the relationship between their resistances. Since the relative changes of resistance caused in the copper coil and the copper shunt by changes in temperature will be the same, their relationship to each other will not change. So the readings of an ammeter connected to a shunt of the same material as the ammeter's coil will not depend on the temperature.

Now let us consider how a voltmeter's readings will be affected by temperature changes.

As you know, voltmeters must have a high internal resistance; for this the instrument's coil, which in general has not got a high resistance, should be connected in series with a high supplementary resistance made of some material with high specific resistance such as manganin or constantan. Therefore the total resistance of the voltmeter is determined basically by this series resistance and if the instrument's readings are not to vary with temperature changes, this series resistance must remain unaltered. Since the resistance of the alloys of high specific resistance mentioned above depend very little on temperature, the readings of a voltmeter are also almost independent in practice of the temperature of the surrounding medium.

Besides its effect on the magnitude of the resistances, change in temperature can also affect the elasticity of the spiral springs which hold the moving parts of the instrument in equilibrium. To obviate this effect the springs of accurate instruments are made from an alloy whose elasticity does not depend on the temperature (elinvar).

260. Since wires $MACN$ and $MBDN$ are connected in parallel between points M and N, the fall of potential is one and the same for the whole length of these wires. The fall of potential is distributed uniformly along each of these wires. Therefore if length MA equals length MB, the potentials at A and B will be the same and current will not pass along wire AB. The same thing can be said of sections MC and MD. If they are of the same length as one another, current will not pass along CD. But wires AB and

CD will be at different potentials and if points E and F upon these wires are connected, current will pass along EF and consequently also along AE, BE, FC and FD. This will happen wherever points E and F are chosen along wires AB and CD, since the potential at E will always be higher than that at F.

261. The amount of heat given off by the current in the rod when the potential difference applied to the ends of the rod is constant will equal, from Joule–Lenz's law,

$$Q = 0 \cdot 24 \, \frac{V^2}{R} t.$$

It will obviously be greater, the lower the resistance R. Since asbestos slows down the giving up of heat to the surrounding atmosphere, the rod will be hotter inside the asbestos. But R for graphite falls as the temperature rises; therefore Q will be greater when the rod is covered with asbestos.

262. Two wires are connected to a plug in a lighting system, neutral and phase. Since the picture on the oscillograph screen did not alter when point A was earthed, it follows that this wire must have been connected to the neutral wire in the plug. The earthing of point B removes the oscillations of the current passing to one pair of the oscillograph's plates and therefore the electron stream traces a line on the screen at right angles to the other pair of plates. The earthing of point C shorts out the phase and neutral wires in the plug, the current in the wire rises sharply and the fuse in the plug burns out.

263. At a given potential at the ends of a circuit the power available in the circuit equals V^2/R, i.e. it is greater, the lower the resistance R. Consequently, in a circuit of 39 bulbs, whose total resistance is less than that of 40, a greater power will be available, therefore the temperature of the heated filaments will be higher and they will give more light. So the room will be lighter from the 39 bulbs than from the 40.

Of course it is not possible to reduce the number of lamps very far simply because they will burn out.

264. The battery E charges the capacitor to a certain potential V, after which the flow of current in C and R_3 stops and continues only in the circuit through R_1 and R_2. When the current in the capacitor circuit stops there will be no fall of potential in resistance R_3 and the capacitor's p.d. must equal that between points a and b, i.e. the fall of potential in resistance R_2. Ohm's law gives us that the current

$$I = \frac{E}{R_1 + R_2}.$$

This current creates a p.d. at the ends of resistance R_2, i.e. between points a and b, of

$$V = IR_2 = \frac{ER_2}{R_1 + R_2}.$$

And this is the potential to which the capacitor C will be charged.

265. The argument leaves out of account one very important factor. As material is deposited at the electrodes of an electrolytic tank polarization of the electrodes takes place, as a result of which a back e.m.f. appears. This phenomenon serves as the basis for the creation of secondary elements, i.e. accumulators. When several cells containing acid solution are connected in series, the back e.m.f., which is the sum of the back e.m.f.'s appearing in each cell reaches the same value as the e.m.f. of the accumulator battery before the quantity of electricity intended has been taken from it and the flow of current will stop.

266. Through any cross-section between the electrodes and at right angles to the direction of motion of the ions, currents pass in the electrolyte of

$$I_+ = en_+v_+, \quad I_- = en_-v_-,$$

where e is the ion charge, n_+ and n_- are the concentrations of the respective ions and v_+ and v_- are their velocities.

The full current is

$$I = I_+ + I_- = e(n_+v_+ + n_-v_-).$$

When n_-v_- negative ions escape, the same number of positive ions remain near the cathode and they, together with the n_+v_+

positive ions which come to the cathode, are deposited there. Thus the number of positive ions deposited at the cathode is determined by the full current.

The position at the anode is similar.

267. The reverse connection would mean that oxygen would be liberated on the rails as a result of electrolysis of the moisture in the track and this would lead to premature corrosion.

268. The electric field in a metal which causes the movement of electrons acts with the same force both on the electrons and on the ions of the metal's structure, but these forces act in opposite directions. The force acting on the electrons causes them to accelerate. Under the electric field's effect, the electrons acquire a certain momentum which they give up to the ions on impact. In this, the average force with which the electrons act upon the conductor when they collide with the ions of the metal's structure is exactly equal to the force with which the electric field acts directly upon the ions, but it is opposite to this force in direction. Therefore a metal conductor through which current is passing does not experience any mechanical forces in the electrons' direction of motion.

269. The velocity of an electron at the anode is determined by the p.d. over the whole distance between cathode and anode and will be greater, the greater is this p.d. If the anode battery has no internal resistance, then the p.d. is the same in both cases and equals the e.m.f. of the battery. In this case the velocity of electrons reaching the anode in arrangement A and arrangement B are the same. If the internal resistance of the battery is considerable, then $V = E - Ir$, where I is the anode current and r is the battery's resistance. Consequently the p.d. at the anode, and so also the velocity of electrons at the anode, will be greater in the case in which the anode current is less. But the electric field at the anode is less in arrangement A than in arrangement B, since that grid is at lower potential. Therefore the anode current is less in arrangement A, and so the velocity of the electrons at the anode will be greater in that arrangement than in B.

270. The A-battery produces a current which passes through the circuit $ABCD$ from the positive to the negative pole. The heater current will equal $5 \div 5 = 1$ A. The electrons move inside the valve from cathode to anode. Therefore the electrons in the external circuit which form the anode current will move towards the cathode from D. Now there are two possible routes for the electrons to take from D to the cathode: DC and DAB. But they meet no resistance on route DC (we have neglected the resistance of wires and ammeters), while on route DAB, their motion is impeded by the e.m.f. of the A-battery (which acts in such a direction that it moves the positive charges from D towards A). Consequently the whole anode current will go along route DC and ammeter 1 will register a current of 1 A, while ammeter 2 will register a current of $1 \cdot 1$ A.

271. For the gas in a neon lamp to shine, an electric field must be created in it. As a result of the friction on the glass tube of the neon lamp electric charges appear and their field causes a brief fluorescence of the lamp.

272. If the needles are drawn aside in opposite directions, their period of oscillation will be somewhat reduced, since the 'arm' of the return force will be increased, i.e. its moment relative to the axis of each needle will be increased.

273. Break the shaving in two and see if the two halves attract one another. If the shaving was magnetized, each half will also be a magnet, having a North and a South pole; these two magnets will react upon each other.

274. It is clear from the drawing (Fig. 240) that it is possible to find, for any edge of the cube, another edge in the corresponding diagonal plane in which there is a current that is equal to and flows

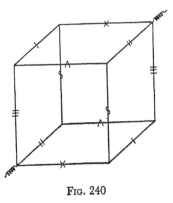

FIG. 240

in the same direction as the current in the first edge. These currents create magnetic fields of equal strength at the centre of the cube but they are opposite in direction; therefore the resultant magnetic field strength will be zero.

275. In the first case the direction of the current in the wire and the filament changes simultaneously. Therefore, depending on how the ends of the filament are connected to the grid, the direction of the currents in wire and filament at any moment will either be the same, in which case the filament will be attracted to the wire, or opposite, in which case the filament will be repelled.

In the second case the direction of currents will be the same for one half-period and opposite for the following half-period. Therefore the filament will be alternately attracted to the wire and repelled, i.e. it will perform oscillations about the position of equilibrium; the frequency of its oscillations will be equal to the frequency of the town-grid's a.c. current.

276. The solenoid creates a magnetic field of a single circular current outside the toroid, its radius approximating to that of the toroid. Therefore the magnetic needle of the compass will be deflected according to the corkscrew rule. In our diagram the needle's North pole will be deflected towards the observer.

277. The magnet will attract the nearest unheated wire of the

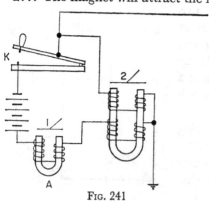

FIG. 241

rotator, but as a result of heating by the burner, this wire will lose its magnetic properties and magnetic attraction between it and the magnet will cease; then the magnet will attract the next wire, which is unheated and therefore strongly magnetized, the process repeating itself indefinitely. As a result there takes place a continuous rotation of the rotator.

278. In Fig. 241 is shown one possible arrangement which allows the reception of copies of telegrams despatched at the sending station and also allows simultaneous transmission of telegrams in both directions. For when key K is depressed, first of all the circuit at sending station A is closed; this circuit contains two electromagnets, 1 and 2, of which only magnet 1 is working, while the other is not, since a current passes through its two coils which create magnetic fields which are opposite in direction. Thus a copy of the telegram being sent is made. Second, a circuit is closed which connects up station B (through earth), where the corresponding electromagnet 2 is working, since only one of its coils is connected to this circuit. It is also clear that when the key is depressed at station B, electromagnet 2 is brought into operation at station A and electromagnet 1 at station B. Thus both stations can operate simultaneously.

279. Unless the steel bar contains residual magnetism, its motion through the copper ring will exercise no effect on the ring's position at all. But when the magnet passes through, then, according to Lenz's law, an induced current appears in the ring which will react on the magnet and impede its movement; consequently the magnet will act upon the ring in its turn with the same force. Therefore the magnet's motion will deflect the ring from its initial vertical position.

280. When the coil is connected to an a.c. circuit, an a.c. current is also induced in the ring, acting at any given moment in the opposite direction to that in the coil, since the ohmic resistance of the ring is very low. The magnetic fields caused by currents flowing in opposite directions react on one another and seek to repel from each other the wires along which these currents are flowing. As a result of the fact that currents flowing in opposite directions repel each other, the ring jumps up. When the coil is connected to a d.c. circuit, the ring will jump away at the moment of connection as a result of the appearance of an induced current acting in the opposite direction.

281. When a magnet falls through a ring an e.m.f. of induction arises in it and an induced current flows through the ring. The interaction of the magnetic fields of this current and the falling magnet are such, according to Lenz's law, that it hinders the motion of the magnet which causes the e.m.f. of induction. Therefore the magnet's fall will take place with an acceleration that is less than that of free fall.

282. Currents will arise in the large copper sheet (Foucault's induction currents), whose magnetic field will significantly, if not entirely weaken the field of coil B, and the voltmeter to which coil A is connected will either register nothing or else only a small voltage.

283. The forces of interaction between two currents are governed by the interaction of their magnetic fields; this interaction can be imagined to be as though every line of force in the magnetic field were seeking to reduce its own length, while neighbouring lines of force acting in the same direction seek to move farther away from each other. The first leads to the currents which act in the same direction being attracted to each other, and the second to the currents which act in opposite directions being repelled by one another. But if coils through which steady currents are passing are passed over a core whose magnetic inductivity is very great, and therefore all the lines of force of the magnetic field pass along the core, then the displacement of either of the coils along the core in no way alters the disposition of the lines of force of the magnetic field inside the core. Insofar as the interaction between current-carrying wires is always linked with a change in their magnetic field, while in the case of a steel core, the character of the magnetic field is set by the form of the core and does not alter when the mutual positions of the coils are altered, there can be no forces of interaction between the coils.

284. The phenomenon of electromagnetic induction consists in the appearance of an electric current in a circuit when the magnetic flux through the circuit is altered. The galvanometer needle connected to the solenoid is deflected only while the magnet is being moved into the coil. When the magnet has been moved in, all change in the magnetic field ceases and, therefore, the induction current ceases too. So in Colladon's experiment the galvanometer needle was deflected while Colladon was moving the magnet into the coil in the first room where he could not see the galvanometer. But when he left the magnet at rest and went to look at the galvanometer, the needle had had time to return to its original position.

285. A steel core with two windings is an ordinary transformer. If one of the transformer's windings is connected to an a.c. source and the voltage at the ends of both windings be measured (V_1 and V_2) with a voltmeter, we can find the relationship between the number of turns in the two windings since

$$\frac{V_1}{V_2} = \frac{n_1}{n_2}, \qquad (1)$$

where n_1 and n_2 are the numbers of turns in the first and second windings respectively. However this method does not allow us to find n_1 and n_2 independently. But if additional windings be wound on to the core with a known number of turns n', and the voltage at the ends of this winding be measured (V'), then we can write down the relationship

$$\frac{V_1}{V'} = \frac{n_1}{n'},$$

from which we can find n_1 and then find n_2 also from equation (1). If the voltmeter is sufficiently sensitive, it is possible to take $n' = 1$, i.e. simply pass a piece of wire across the yoke of the core and connect the ends to the voltmeter's terminals.

286. Two exactly identical windings wound on to the same core in the same direction and connected in parallel are equivalent to one winding of wire with cross-section twice as large. Hence the resistance is halved while its inductance remains practically unchanged. But inasmuch as the ohmic resistance is so small by comparison with the inductive resistance, the total resistance of the two windings will remain unaltered and the current will not alter.

287. In the first case the work expended on displacing the conductor is entirely converted into heat, given off in resistance R; in the second case part of the work expended goes on increasing the magnetic energy of the field which arises round the inductor.

The work expended on displacing the conductor in unit time equals EI, where E is the e.m.f., identical in both cases, arising as the result of the conductor's displacement in the magnetic field, and I is the current in the circuit. Current I_1 in the first case is greater than current I_2 in the second case, since the inductor's e.m.f. retards the velocity of increase of the current. Consequently too, work in unit time $EI_1 > EI_2$.

This is also true for uniform motion of the conductor, but only up to the establishment of steady current in the circuit, after which the presence of self-induction in the circuit does not have any effect.

XI. Optics

288. The reason is that the source of light (the candle flame) is drawn out in a vertical direction. When the fork is set vertically, the boundary between light and shadow for each of the prongs on the screen is in approximately the same place for all points of illumination from the source, and therefore the shadow of the prongs is distinct. But when the fork is set horizontally, then the boundary between light and shade made by one point in the source of light with a given prong will be displaced on the screen relative to the boundary between light and shade made with the same prong by another point in the light source; and so the whole shadow cast by the fork will be blurred.

289. The addition of a mirror has the same effect as the addition of a second, virtual, source of light S, placed at a distance of $3a$ from the screen (a virtual image S' of the source is obtained behind the mirror and at a distance of a from it (Fig. 242)). We know that the illumination from a point source of light is directly proportional to the strength of the light and inversely proportional to the square of the distance. If the original source of

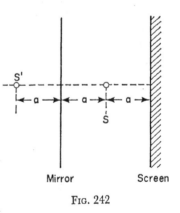

Fig. 242

light gives an illumination at the centre of the screen of I/a^2, then the second source adds an illumination of $I/(3a)^2$. Therefore if a mirror is added, the illumination at the centre of the screen will be

$$E = \frac{I}{a^2} + \frac{I}{9a^2} = \frac{10}{9} \cdot \frac{I}{a^2} \cdot$$

290. Let us construct the image of an object AB in a plane mirror CD which is placed parallel to the object (Fig. 243). We

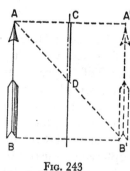

know from the laws of reflection of light that the image $A'B'$ in the plane mirror will be located symmetrically to the object AB, i.e. at the same distance from the mirror as the object ($CA' = CA$). It is clear from the construction that it is enough to have a mirror of such a height (CD) that points A' and B' can both be seen from point A. But $CD = \frac{1}{2}A'B' = \frac{1}{2}AB$, i.e. it is enough to have a mirror half as high as the man is tall.

Fig. 243

291. Since the clouds floating in the sky are at a greater distance from the camera-lens than the man, the image of the clouds will lie nearer to the lens than that of the man. For their image to fall upon the film, the extension of the camera must therefore be decreased.

292. The number of images will be infinite. The first image O_1 of point O in mirror M_1 produces a reflection O'_2 in mirror M_2;

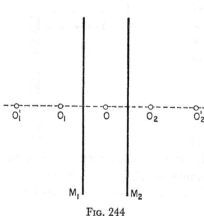

in its turn O'_2 is reflected in M_1 and so on (Fig. 244). The reflection of point O in mirror M_2 will produce a similar sequence of images. In this way an infinite series of images is caused at distances of x from each other, x being equal to the distance between the mirrors.

Fig. 244

Of course, with every

succeeding reflection the brilliance of the image is weakened and so in practice the number of images will be finite.

293. When a point of light A is reflected in mirrors OM and OM' two images are formed, A_1 and A_2. The position of each of

these can be found in the usual way, i.e. the perpendicular is dropped from the point of light on to the plane of the mirror and then produced to the same distance beyond the plane of the mirror. But in addition to this we must consider the possibility of an image being formed of the virtual source A_1 in mirror OM' and, similarly, of A_2 in mirror OM (Fig. 245).

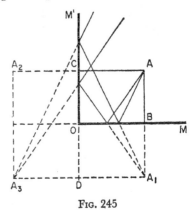

Fig. 245

Let us find the image A_3 of the source A_1 in mirror OM'. It will appear to the eye that this image lies at the same distance behind mirror OM' as the virtual source A_1 lies in front of this mirror. Therefore, to find the position of A_3, we must drop the perpendicular from A_1 on to the line OM' produced and then produce this perpendicular a distance $A_1D = DA_3$.

The virtual source A_1 cannot be reflected in mirror OM, since it lies behind the plane of this mirror. Similarly, there will be no image of the virtual source A_2 in mirror $OM.'$

It still remains to find the image of the virtual source A_2 in mirror OM. But if the above construction be repeated for this case, it is not hard to see that the image of the virtual source A_2 in mirror OM and that of the virtual source A_1 in mirror OM' coincide.

Thus three images in all are to be seen: A_1, A_2, A_3.

294. The mirrors should be arranged at right angles to one another, as in Fig. 246. Then

$$\alpha + \delta = \beta + \gamma = 90°,$$

and therefore

$$\epsilon + \theta = 360° - (\alpha + \beta + \gamma + \delta) = 180°$$

It is not hard to see that this equation holds good whatever the angle of incidence.

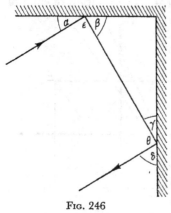

295. The solution to this problem follows from that of the previous one: the mirrors should be so arranged that they are at right angles to one another and meet at one vertex. Such a hollow pyramid, with three reflecting sides at right angles to one another will reflect any ray in a line parallel to the line of incidence.

296. Light can be diffused in a curve in a medium with a variable refractive index. Consider the path of a ray through a series of

FIG. 246

plane parallel laminas (Fig. 247), with gradually changing refractive index

$$n_0, \, n_1, \, n_2, \, n_3, \, n_4, \, \ldots \text{ and so on.}$$

Let

$$n_0 < n_1 < n_2 < n_3 < n_4 \ldots \tag{1}$$

As it passes from one lamina to another, the ray of light changes its direction each time:

$$\frac{\sin i_1}{\sin r_1} = \frac{n_1}{n_0}, \quad \sin r_1 = \frac{n_0}{n_1} \sin i_1.$$

By construction it is clear that $\sin r_1 = \sin i_2$.

When the ray is refracted in the next lamina

$$\frac{\sin i_2}{\sin r_2} = \frac{\sin r_1}{\sin r_2} = \frac{n_2}{n_1},$$

hence

$$\sin r_2 = \frac{n_1}{n_2} \sin r_1 = \frac{n_0}{n_2} \sin i_1.$$

Further $\sin r_2 = \sin i_3$;

$$\frac{\sin i_3}{\sin r_3} = \frac{\sin r_2}{\sin r_3} = \frac{n_3}{n_2};$$

$$\sin r_3 = \frac{n_2}{n_3}\sin r_2 = \frac{n_0 n_2}{n_2 n_3}\sin i_1,$$

$\sin r_3$ being equal to $\sin i_4$.

The next refraction gives us:

$$\frac{\sin i_4}{\sin r_4} = \frac{\sin r_3}{\sin r_4} = \frac{n_4}{n_3}$$

from which $\sin r_4 =$

$$\frac{n_3}{n_4}\sin r_3 = \frac{n_0}{n_4}\sin i_1.$$

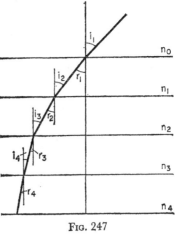

Fig. 247

On the strength of equation (1), it is clear that $\sin r_1 > \sin r_2 > \sin r_3 > \sin r_4$, i.e. the ray of light changes its direction every time it passes from one lamina into another. If the refractive index gradually rises the angle of refraction correspondingly decreases. If the ray of light passes through a large number of such thin laminas with a gradually changing refractive index, the ray's path will be curved.

Media of this kind, with gradually changing refractive index, are for example, non-homogeneous liquids or layers of the atmosphere of gradually changing density.

297. The observer will first see the sun's rays after the period of time has elapsed which is necessary for light to travel the distance from the moon to the observer's eye, i.e. approximately $\frac{1}{3}$ sec.

298. Refraction in the earth's atmosphere of the sun's rays which come from the ends of the diameter or the chords of the sun's disk which are parallel to the plane of the visible horizon takes place under identical conditions and therefore we do not notice any distortion of the sun's disk in the horizontal plane. But it is a different matter with the rays which come from the ends of the diameter and the chords of the sun's disk which are

238

perpendicular to the plane of the visible horizon. When an observer sees the lower edge of the sun's disk, the rays from the lower edge strike the upper edge of the earth's atmosphere at a greater angle than the rays which come from the upper edges of the sun's disk and therefore they are bent more sharply than the

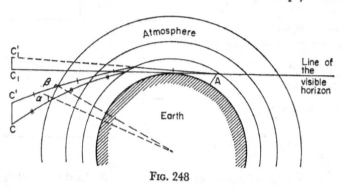

FIG. 248

latter. Therefore the lower edge of the sun's disk appears to us to be raised higher than the upper edge is raised (see Fig. 248, where A is the observer, CC' is the diameter of the sun perpendicular to the plane of the visible horizon and $\alpha > \beta$). This results in the sun appearing to be flattened in this plane.

299. A and B will see each other first at the moment when the extreme ray of light from A strikes B after being reflected from the mirror and, conversely, when a ray of light from B strikes A (Fig. 249). Since the angle of incidence equals the angle of reflection, it is evident that rays BO and AO must strike the mirror at the same angle.

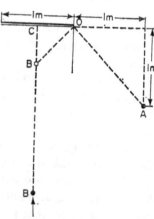

FIG. 249

Therefore with the measurements of the mirror given and the given position of A, we must have

$$BC = CO = \tfrac{1}{2}\,\text{m},$$

i.e. A and B will catch sight of each other when B is at a distance of $\tfrac{1}{2}$ m from the mirror.

300. Let us take point A on the edge of the sun (S) and construct the path of the rays from A which strike the small mirror lying on the table and are reflected from it (Fig. 250). The beam

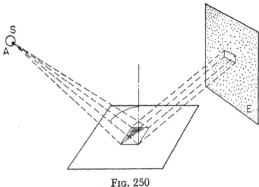

Fig. 250

of rays which comes from this point can be considered to be parallel. In this case the rays will all strike the mirror at the same angle, and will therefore all be reflected from the mirror at the same angle; thus, a point A on the sun will produce an illuminated patch on a screen which will in general have the form of a quadrilateral; the exact shape of this quadrilateral will depend on the angle at which the pencil of rays from A falls upon the mirror. (If this angle equals 45°, then both sides of the mirror's square will be projected on to the screen E as equal sides.) But the sun has finite dimensions. Therefore the beams proceeding from different points on the sun will produce quadrilateral patches of light which are not directly superimposed on one another on the screen and the sum result of these patches will be a patch of elliptical (or almost circular) shape. This will be true if the screen is at a sufficient distance from the mirror. If the screen is near the mirror, then, as a result of the narrowness of the conical beam proceeding from the sun, the patches will be so superimposed on one another that the resulting patch is either a rectangle or else a square, with slightly blurred edges.

301. For the eye's focal length to remain the same both in the air and under water, there must be no refraction of rays proceeding from distant objects (i.e. parallel rays) on the front surface of the cornea. Therefore this surface must be flat.

302. Let us construct the image of an object AB, obtained on a screen by using lens L (Fig. 251). To do this it is usual to draw

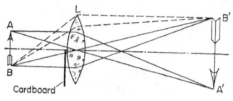

FIG. 251

two rays from each of the extremities A and B of the object, the path of these rays being known (one of them being parallel to the principal axis, and the other passing through the lens's optical centre). But it is clear that at point B' (for example) meet not only the two rays which leave 'B already mentioned, but the whole cone of rays which proceeds from B and strikes the lens (two of these rays are shown in the figure as dotted lines). Therefore if half the lens is covered by a piece of cardboard, rays will still strike the screen from every point of the object, i.e. in this case too a full image of the object will be obtained; but the amount of rays will be twice as few and therefore the image will be half as bright.

303. To obtain an image on the screen of the source of light the distance between the source and the screen must be increased by quite simple means, without changing the position of either. This can be done by putting a plane mirror at a suitable distance behind the source. So that the light from the source it-self should not hinder this, it can be screened with a small opaque screen.

304. To obtain a real image in the air with the help of a double concave lens, a convergent beam must be directed on to it. This can be done by placing in front of the double concave lens, i.e.

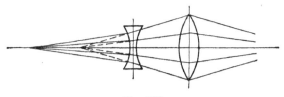

FIG. 252

nearer the source of light, a converging lens of suitable focal length (Fig. 252).

305. When a ray of light strikes the sphere at an angle of α (Fig. 253), it is refracted at an angle of β and enters the sphere. Is is clear from elementary geometrical considerations that the ray strikes the surface of the sphere at A at an angle of β. Therefore it is of necessity refracted and leaves the sphere at an angle of α (the reversibility of the path of rays of light). Of course, a small part of the light will be reflected back inside

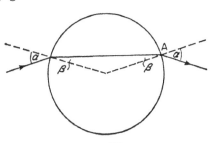

FIG. 253

the sphere at point A and the ray will leave the sphere somewhat weaker.

306. For our construction we shall make use of two basic properties of a lens:

(1) A ray of light that passes through the optical centre of a lens, passes through the lens without being refracted.

(2) All the rays which strike a lens in a parallel beam converge at one point after refraction in the focal plane of the lens.

Draw the focal plane through point F. Ray BC cuts it at point

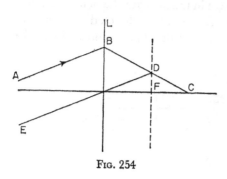

D (Fig. 254). Draw ray ED, passing through the optical centre of the lens O and also through D. On the basis of the two properties of the lens mentioned above, we draw the conclusion that before refraction in the lens, the unknown ray of light must be parallel to ray ED and strike the lens at

Fig. 254

point B. Thus the path of ray BC before it is refracted in the lens will be represented by the line AB.

307. From the fact that object and image are located on opposite sides of the optical axis, we can deduce that the image is

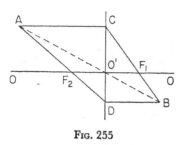

real and obviously obtained with the help of a double convex lens. Therefore the line drawn from A to B will cut the optical axis of the lens OO in O', its optical centre (Fig. 255). Erect the perpendicular to OO at O' and draw ray AC from A parallel to the optical axis. After refraction in the lens, this ray must pass through the

Fig. 255

second principal focus of the lens F_1 and strike point B. The point of intersection of CB and OO' also gives the location of focus F_1. By drawing lines BD and DA we find the first principal focus F_2 in a similar way.

308. The lenses should be arranged so that their principal foci coincide. The path of the rays is shown in Fig. 256.

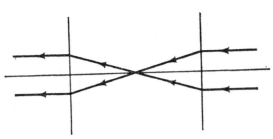

Fig. 256

309. Let us suppose that the source of light S lies at a distance from the converging lens which is twice that of the focal length (Fig. 257a). Then the cone of rays proceeding from S and striking the lens will converge beyond the lens at S', which lies at a distance of more than F and less than $2F$ from the lens. The

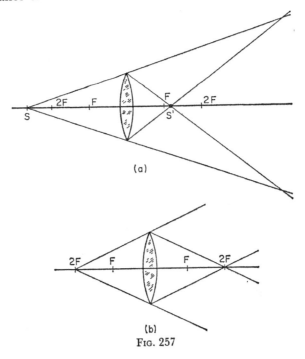

(a)

(b)

Fig. 257

cone of rays striking the lens is wider than the cone of rays bypassing the lens. Therefore there will be a region in space in
which the rays, as they diverge from image S' will cut the rays
proceeding from S and bypassing the lens. It is clear that at any
point inside this region it is possible to see both the source S and
its image S' at the same time. To satisfy the requirements laid
down by the conditions of the problem it is clearly essential for
the cone of rays converging beyond the lens at S' to be narrower
than the cone of rays proceeding from S and striking the lens.
In the borderline case both cones must be of the same width
(Fig. 257b). Then the extreme rays of these two cones will not
intersect at any point and there will be no points in space from
which S and S' can be seen simultaneously.

310. A parallel beam of rays remains parallel after striking and
passing through a disk with parallel plane surfaces. Imagine the

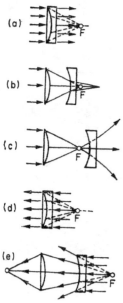

FIG. 258

disk cut as indicated in the problem,
but leaving the lenses together (Fig.
258a). Then everything remains as it
was. But we can look upon this case in
the following way: when the rays fall
from the side of the converging lens they
converge at the lens's focus F, which is
also the virtual focus of the diverging
lens and therefore these rays leave the
diverging lens in parallel lines. If the
lenses be moved apart a little, then the
focus F, at which the rays ought to meet
after passing through the converging lens,
will be nearer than the virtual focus of
the diverging lens. Therefore the rays
will leave the converging lens in converging lines (Fig. 258b). Evidently this
result depends on the distance through
which the lenses are moved apart. If this
distance is greater than the focal length,

then the rays will strike the diverging lens in a diverging beam and will diverge still further after passing through this lens (Fig. 258c).

Now let us consider the case when a beam of parallel rays falls upon the lenses while still in contact, but from the side of the diverging lens. We can look upon this case in the following way: after passing through the diverging lens, the beam of parallel rays converges at the virtual focus F, which is at the same time the focus of the converging lens. Therefore the rays will leave the converging lens in parallel lines (Fig. 258d). If the lenses are moved apart, the virtual focus F of the diverging lens will be further away from the converging lens than its own focus, and therefore the rays will leave the converging lens in converging lines (Fig. 258d). Clearly this result will not depend on the distance through which the lenses are moved apart.

311. The rays of different colour can be separated by placing the disk at an angle to the beam and by selecting this angle in such a way that the beam's angle of incidence lies between the critical angle for total internal reflection for the two colours. Then the rays of one colour will be reflected to one side, while those of the other will pass through the disk (Fig. 259).

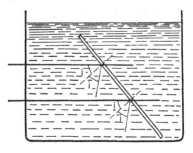

FIG. 259

312. For a man to be able to see, light must be absorbed by the retina of his eye. But if a man became completely transparent, then his retina is also transparent and so cannot absorb light. Besides, for a man to be able to see (and not just sense light) an image of objects must be formed on the retina. Therefore, if the membranes of his eye are transparent, the man will also lose the ability to see images, since the light will strike the retina besides the pupil.

313. Because the reflection of light is always greater from walls than that from transparent windows, i.e. objects which allow the light to pass through.

314. The focal length of the eye, as of any lens, is different for different wavelengths, i.e. for different colours of the spectrum. Red rays are refracted less violently and therefore the visual impression is caused that red objects are nearer to the observer than blue ones (Fig. 260).

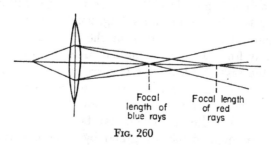

Focal length of blue rays

Focal length of red rays

FIG. 260

315. Damp sand appears to be dark because reflection from the sand has been considerably reduced and a great part of the light passes inside the sand, where it is absorbed.

316. At the temperature of a stove the greatest quantity of energy falls within the infra-red range of the spectrum and this is largely kept in by glass. But the greatest quantity of energy in the sun's spectrum falls within the visible part of the spectrum, which passes through glass. Therefore by covering a greenhouse with glass we allow the heat of the sun to be transmitted to the ground without allowing the heat of the ground to be lost outside.

317. No, since radiation of energy from the sphere takes place at the same time as absorption of energy by it. Radiation increases very rapidly as the sphere's temperature rises. So eventually there will come a moment when the sphere will radiate as much heat as it receives. Then the sphere's temperature will cease to rise. What temperature is established depends on the relationship between the dimensions of the mirror and the hole in the sphere.